# Dynamics of Molecules and Materials-II 2013



Journal of Physics: Conference Series Volume 554

## Glasgow, United Kingdom
## 5-6 July 2013

**Editors:**

**Monica Jimenez-Ruiz** **Stewart F. Parker**

ISBN: 978-1-63439-597-7
ISSN: 1742-6588

**Printed from e-media with permission by:**

Curran Associates, Inc.
57 Morehouse Lane
Red Hook, NY 12571

**Some format issues inherent in the e-media version may also appear in this print version.**



Printed by Curran Associates, Inc. (2015)



Institute of Physics
Dirac House, Temple Back
Bristol BS1 6BE UK

Phone: 44 1 17 929 7481
Fax: 44 1 17 920 0979

techtracking@iop.org

**Additional copies of this publication are available from:**

Curran Associates, Inc.
57 Morehouse Lane
Red Hook, NY 12571 USA
Phone: 845-758-0400
Fax: 845-758-2634
Email: curran@proceedings.com
Web: www.proceedings.com

# TABLE OF CONTENTS


**012001 IN1-LAGRANVE - The New Ill Instrument to Explore Vibration Dynamics of Complex Materials** ........ 1
*A Ivanov, M Jimenéz-Ruiz, J Kulda*

**012002 Parameters of the NERA Spectrometer for Cold and Thermal Moderators of the IBR-2 Pulsed Reactor** ........ 8
*I Natkaniec, D Chudoba, L Hetmanczyk, V Yu Kazimirov, J Krawczyk, I L Sashin, S Zalewski*

**012003 Recent and Future Developments on TOSCA at ISIS** ........ 20
*S F Parker, F Fernandez-Alonso, A J Ramirez-Cuesta, J Tomkinson, S Rudic, R S Pinna, G Gorini, J Fernández Castañon*

**012004 The Use of Direct Geometry Spectrometers in Molecular Spectroscopy** ........ 29
*S F Parker, A J Ramirez-Cuesta, P W Albers, D Lennon*

**012005 Sample Environment Issues Relevant to the Acquisition of Inelastic Neutron Scattering Measurements of Heterogeneous Catalyst Samples** ........ 38
*R Warringham, D Bellaire, S F Parker, J Taylor, R A Ewings, C M Goodway, M Kibble, S R Wakefield, M Jura, M P Dudman, R P Tooze, P B Webb, D Lennon*

**012006 Hydrogen Gas Sample Environment for TOSCA** ........ 50
*M G Kibble, A J Ramirez-Cuesta, C M Goodway, B E Evans, O Kirichek*

**012007 Cryogenic Sample Environment on TOSCA** ........ 57
*R B E Down, A J Ramirez-Cuesta, R A Major, J Keeping, S Rudic, O Kirichek*

**012008 Vibrational Spectra of Ferrocene, Ferrocene-containing Polymers and Their Oxidized Compounds** ........ 63
*M Appel, B Frick, A Ivanov, J Elbert, M Rehahn, M Gallei, T L Spehr, B Stühn*

**012009 Hydrogen on Graphene Investigated by Inelastic Neutron Scattering** ........ 71
*C Cavallari, D Pontiroli, M Jiménez-Ruiz, A Ivanov, M Mazzani, M Gaboardi, M Aramini, M Brunelli, M Riccò, S Rols*

**012010 An Inelastic Incoherent Neutron Scattering Study of the Internal Dynamics of Ergocalciferol and Cholesterol** ........ 78
*K Holderna-Natkaniec, I Natkaniec, Z Fojud, B Peplinska*

**012011 Ph-induced Size Changes in Solutions of Cholesteric Liquid- Crystal Polymers Studied by SANS** ........ 94
*S M Pérez Méndez, D Rodriguez Martinez, S M King*

**Author Index**

# Dynamics of Molecules and Materials - II

This meeting was the second conference on the Dynamics of Molecules and Materials (DMM-II). The meeting was held at University of Glasgow on the 5th and 6th July 2013. The aim of the meeting was to follow a successful meeting at the Institut Laue Langevin (Grenoble, France) in 2007 that brought together people that were interested in, or involved in, the use of inelastic neutron scattering spectroscopy to study problems in chemistry and materials science.

The six sessions covered topics from catalysis to energy materials to polymers and biology. There was a very strong session on instrumentation which provided an up-to-date view of the most important developments in the field of neutron spectroscopy. It had not been the original intention to publish the proceedings but it was agreed that the meeting had generated a unique and valuable overview of the state-of-the-art of neutron spectroscopy in instrumentation, sample environment equipment and applications that a permanent record of the meeting was worthwhile.

Much of the hard work of organising the venue, dealing with accommodation and registration was carried out by Robbie Warringham (University of Glasgow) and Laurence Tellier (ILL) and we thank them for their tireless efforts that ensured the smooth running of the meeting. The conference would not have been possible without the generous sponsorship provided by the Institut Laue Langevin (Grenoble, France), the STFC ISIS Facility (Chilton, UK) and the University of Glasgow. The organizing committee are very grateful to these organisations for their support.

*Organizing Committee*
Victoria Garcia-Sakai (STFC ISIS Facility, UK)
Alexandre Ivanov (ILL, France)
Monica Jiménez-Ruiz (ILL, France)
Mark R. Johnson (ILL, France)
David Lennon (University of Glasgow, UK)
Timmy Ramirez-Cuesta (STFC ISIS Facility, UK)

*Proceedings Editors*
Monica Jiménez-Ruiz (ILL, France)
Stewart F. Parker (STFC ISIS Facility, UK)

NEUTRONS FOR SCIENCE®

Science & Technology Facilities Council ISIS

VIA VERITAS VITA
UNIVERSITY of GLASGOW

Journal of Physics: Conference Series **554** (2014) 011002 doi:10.1088/1742-6596/554/1/011002

# Peer review statement

All papers published in this volume of *Journal of Physics: Conference Series* have been peer reviewed through processes administered by the proceedings Editors. Reviews were conducted by expert referees to the professional and scientific standards expected of a proceedings journal published by IOP Publishing.

Dynamics of Molecules and Materials-II IOP Publishing
Journal of Physics: Conference Series **554** (2014) 012001 doi:10.1088/1742-6596/554/1/012001

# IN1-LAGRANGE - the new ILL instrument to explore vibration dynamics of complex materials

**Alexandre Ivanov, Mónica Jiménez-Ruiz and Jiri Kulda**

Institut Laue - Langevin, 71 avenue des Martyrs, 38000 Grenoble, France

aivanov@ill.fr

**Abstract**. We report on the first experimental results obtained with the recently commissioned IN1-Lagrange spectrometer, which has replaced the old IN1 Be-filter setup used for studies of lattice dynamics and molecular vibrations in an extended energy range, typically used for materials containing light chemical elements, in particular hydrogen. Unlike its predecessor, IN1-Lagrange uses, in addition to a cooled Be-filter, a large area crystal analyzer to collect the scattered neutrons and to define their final energy of 4.5 meV. Using the available set of double-focussing monochromators the incident neutron energy on IN1 can be scanned over the entire range of 4 – 600 meV provided by the ILL hot neutron source. Thanks to its large solid angle of 2.5 steradian and flexibly tuned energy resolution in the range of 1 - 5%, the IN1-Lagrange setup offers an unprecedented sensitivity permitting very rapid data acquisition and revealing fine spectral features in 'standard' samples as well as efficient measurements of hydrogen vibrations in samples containing equivalent masses of $H_2$ as low as 100 μg.

## 1. Introduction

Neutron scattering is an extremely powerful technique in the study of atomic and molecular vibrations. Its sensitivity to all types of atomic movements makes neutron spectroscopy the experimental method of choice when detailed knowledge of the vibration spectrum of a material is requested – on the level of phonon dispersion in single crystals or vibrational density of states (VDOS) in polycrystalline samples. In spite of the integral character of the VDOS, it represents an invaluable input for calculating thermodynamic properties and evaluating forces acting between atoms and molecules constituting condensed matter. It is particularly important in the studies of new and applications related materials typically available in polycrystalline form only.

A number of neutron spectrometers suitable for VDOS measurements are available at existing neutron sources, both pulsed and steady state. For the measurements of VDOS in a large range of energy transfer, typically indirect geometry time-of-flight (TOF) spectrometers (see, for instance, [1]) are exploited such as TOSCA [2] at ISIS (Oxford, UK), which has become a benchmark instrument in the matter, or NERA [3] in JINR (Dubna, Russia). Further development of the TOSCA-type set-up is represented by the spectrometer VISION [4] recently commissioned at SNS (Oak-Ridge, USA). Instruments of this type use Bragg reflection to select a scattered neutron energy close to the wavelength cut-off edge of a beryllium filter, used to suppress higher order harmonics. They provide suitable energy resolution, monotonous and slowly varied as a function of vibration frequency, in the whole range of measurements. The other examples of filter-based indirect geometry spectrometers dedicated to vibrational spectroscopy are the TOF spectrometer FDS [5] at LANSCE (Los-Alamos, USA) and the reactor-based FANS [6] at NIST (Gaithersburg, USA).

Direct geometry chopper spectrometers (see also in [1]) with a large solid angle for scattered neutrons can be used in the VDOS studies on polycrystals too as they may be advantageous in certain energy ranges, depending on the incident neutron wavelength. However, if a broad energy range is to be sampled with more than an order of magnitude difference between minimal and maximal measured frequencies in the spectrum, as it is typical for vibrational spectroscopy, direct geometry instruments would require recording spectra with several incident neutron energies in order to provide suitable resolution conditions in the full range with the need for subsequent normalization.

In the following we describe another type of indirect geometry spectrometer, which combines the advantages given by the continuous beam from a reactor neutron source and a large accepted solid angle for neutrons scattered by the sample. The new instrument with the secondary spectrometer named IN1-Lagrange is installed at the hot neutron source of the high flux ILL reactor [7]. It replaces the former filter-analyser spectrometer (IN1-BeFilter) which had a limited scattering angle and modest energy resolution. The new instrument fully benefits from the new large double-focussing multi-face crystal monochromator supplying monochromatic beam in the intermediate and high neutron energy ranges (mosaic Cu220 and Cu331 crystal faces) as well as in the lower energy range (elastically bent Si111 and Si311 reflections).

## 2. Instrument description

The IN1-LAGRANGE (LArge GRaphite ANalyser for Genuine Excitations) secondary spectrometer setup (Fig. 1) is based on the space focussing of neutrons scattered by the sample in a very large solid angle, which are all then recorded with a relatively small single counter (a $He^3$ gas detector). That is, contrary to typical TOF instruments, the increased solid angle is not accompanied by an increase in the counting volume and the corresponding background.

The focussing reflecting surface of ~1 $m^2$ is built around the vertical sample-detector axis from pyrolytic graphite (PG) crystals set to reflect neutrons with the fixed average energy of 4.5 meV. Altogether there are 612 crystal holders mounted on the solid mechanical structure so that the individual flat crystals are set tangentially to the calculated ellipsoid of rotation. The crystal sizes, their mosaic spread and the height of the part covered by PG crystals are optimised in a way such that all three corresponding contributions to the energy resolution are equal. It is important that the requested energy resolution of the instrument is achieved with a relatively low grade of the pyrolytic graphite.

The appropriately shaped beryllium filter is installed immediately after the sample in order to remove higher-order harmonics in the analyser reflections. The filter transmission below the cut-off is enhanced by cooling it to the base temperature (20 K) of a powerful closed-cycle helium refrigerator.

The resulting accepted solid angle ($\sim 0.8\pi \approx 2.5$ Steradian) is among the highest apertures open for scattered neutrons in the most ambitious instrument projects at present. At the same time it is realised within the highly compact characteristic spectrometer volume, about 1 $m^3$ for the analyser + filter pumped volumes).

The carefully designed screen of boron-containing absorber is installed on the sample-detector axis in order to suppress the intense elastic scattering from the sample. Further reduction of the instrument background, contaminated with the high neutron energy components, is achieved through massive polyethylene shielding built around the whole analyser.

The whole instrument can be considered as a variant of a classical three-axis spectrometer with a very large barrel-like fixed-energy analyser. The major difference is in the changed trajectory for scattered neutrons: if the incident monochromatic beam is kept in the horizontal plane, the scattered neutrons propagate in the directions out of this plane, all around the vertical axis sample-detector. Such arrangement appears to be more convenient for positioning and handling the sample environment equipment (sample cryostat *etc*.).

Note that there are no mutually moving parts within the secondary spectrometer which is positioned as a whole around the monochromator in order to record the inelastic scattering spectra changing step-by-step the incident energy similarly to a typical experiment on a three-axis spectrometer. The energy transfer (it is always, at IN1-Lagrange, a neutron energy loss in the scattering), which corresponds to excitation energy in the sample, is obtained by a subtraction of the analyser energy of 4.5 meV from the energy of incident monochromatic neutrons.

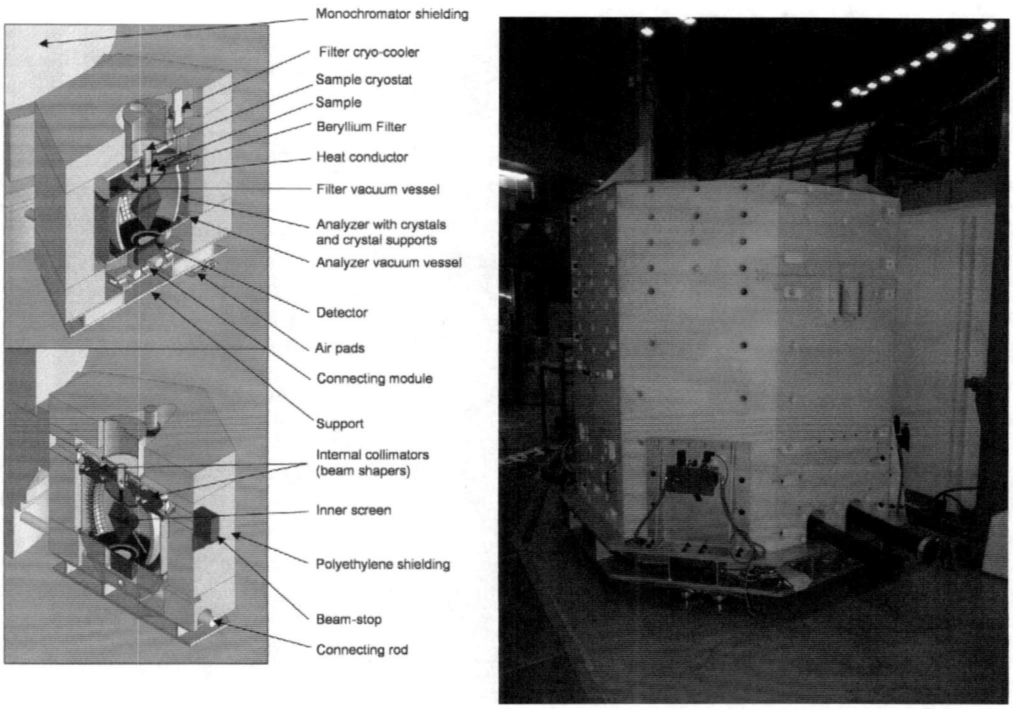


**Figure 1.** Left panel : the IN1-Lagrange analyzer scheme, two different vertical cuts; right panel : the IN1-Lagrange spectrometer assembled for the first tests with neutrons.

## 3. Results of the instrument commissioning

The first inelastic scattering spectrum measured with IN1-Lagrange is presented in Fig. 2. For characterisation of the instrument performance a sample of 2,5-diiodothiophene, $(CH)_2(CI)_2S$, has been used. This material is known to exhibit very sharp vibration lines at low temperatures so that it can be used as a standard for measurements of the spectrometer resolution [8]. The visible sharp peaks in the energy spectrum in Fig. 2 are due to vibrations involving hydrogen atoms. These atoms are present in a relatively small quantity: the hydrogen partial weight in the measured sample amounts to 6 milligram. The collected statistics in the energy interval below 200 meV, left panel on the Fig.2, correspond to some 12-30 seconds measuring time per point, so that this informative energy interval could be scanned in less than 2 hours while the whole scan time for the shown spectrum, up to 600 meV (both panels of Fig. 2), is about 4 hours. The observed about an order of magnitude gain in the count rate on IN1-Lagrange with respect to the IN1 Be-filter setup is found to be in a very good agreement with the project predictions.

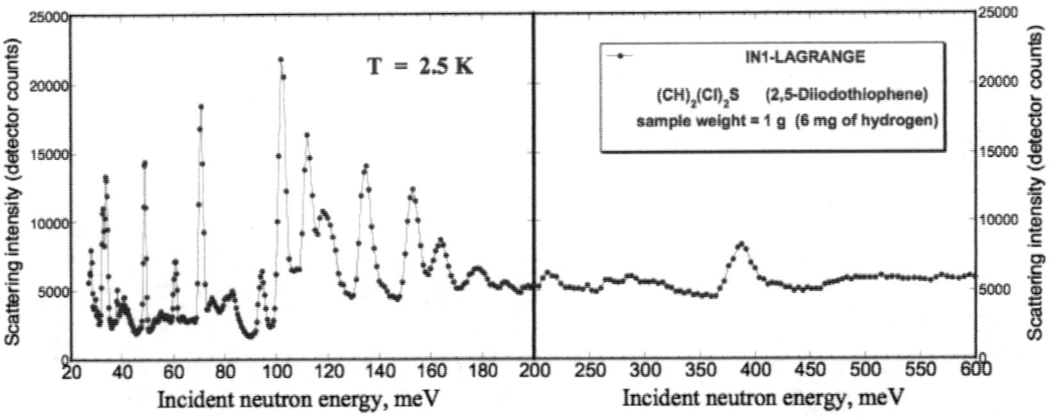


**Figure 2.** The inelastic scattering spectrum recorded with the IN1-Lagrange analyser. Note the different energy scales of the two panels. The sample of 2,5-diiodothiophene is cooled to the cryostat base temperature of 2.5 K. The measurement has been performed in a "fixed-monitor" regime with the 2D-focussing Cu220 monochromator reflection. The intensity is as-measured, without any further treatment, no background (cryostat and sample holder) has been subtracted.

The energy resolution of IN1-Lagrange can be further improved in the broad energy range using the available Cu331 reflecting plane of the IN1 monochromator and introducing fine collimators in the monochromatic beam. As it is seen from the spectra presented in Fig. 3, many more fine details of the vibrational density of states are revealed in the higher resolution run with the Cu331 face at the expense of a solely ~2-fold decrease in the detector count rate. The change-over between different resolution settings can be done on IN1-Lagrange in less than half-an-hour thus offering very flexible conditions for optimising measurements with respect to the resolution *vs* intensity choice.

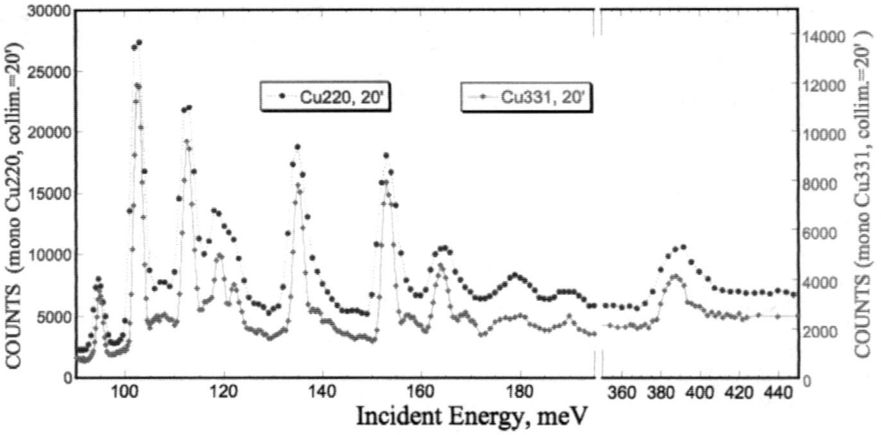


**Figure 3.** A fragment of the 2,5-diiodothiophene vibrational spectrum measured with different monochromator reflections, Cu220 and Cu331, and a 20' slit collimator in the monochromatic beam. No background has been subtracted. Notice the change in the horizontal axis scale in the different energy ranges. The counting time per point, for the chosen monitor counts, was about 20-30% higher in the case of the Cu220 monochromator what corresponds, with the presented intensity scales, to about a factor of 2 intensity decrease, if normalized to the same counting time, for the high-resolution setup with the Cu331 reflection.

The measured line-widths with the sample of 2,5-diiodothiophene are presented as points in the Fig. 4 which also shows calculated energy resolution curves with different analyser and monochromator settings. Similarly to the predictions of the intensity gain, the resolution performance is in full agreement with what was expected from the pre-project estimations.

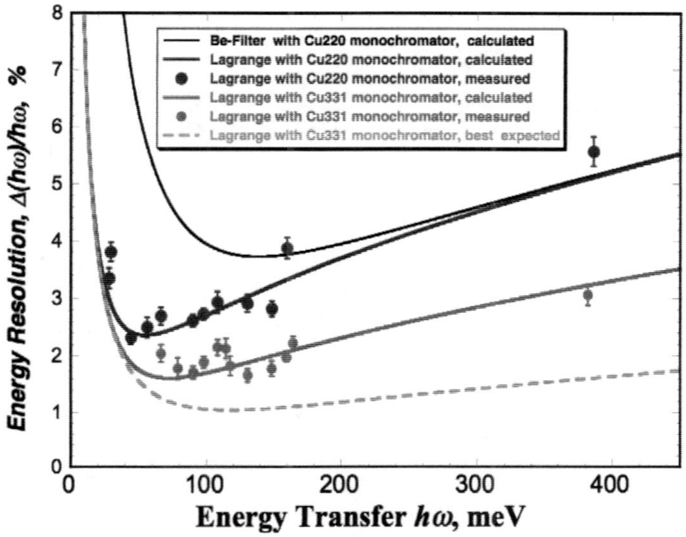


**Figure 4.** Calculated (lines) and measured (points) relative energy resolution of IN1-Lagrange with different primary spectrometer settings. No adjustable parameters are used in calculations which are based on the geometry of the incident and the scattered beam paths. The black line corresponds to the Be-Filter analyser resolution with the Cu220 monochromator, blue line –IN1-Lagrange with the Cu220 monochromator, red line – IN1-Lagrange with the Cu331 monochromator. These calculations are made for the natural collimation of the 'in-pile' beam tube of IN1 and 20' Soller collimator in the monochromatic beam. Points are the peak widths divided by the peak energies as measured at the corresponding conditions with the 2,5-diiodothiophene sample. The scatter of points is most probably due to an uncertainty in the local background close to a line while fitting scattering intensity profiles to Gaussian line shapes. The lowest broken line represents the best-expected future energy resolution of the IN1-Lagrange. It is achieved with about 4-5 fold decrease in peak intensity as compared to the presented Cu331 data.

Further improvements in the resolution power of IN1-Lagrange are envisaged in the near future when the collimator changer will be installed before the monochromator, the work is awaiting the planned change of the reactor beam tube. The best energy resolution, achieved with 10' slit collimators before and after the monochromator, will approach 1% in the energy range 100-200 meV as shown in Fig. 4. This ultimate energy resolution, positioning IN1-Lagrange at the top of high-resolution neutron spectrometers worldwide, will bring new motivation for progress in theoretical model and *ab-initio* calculations for existing and future functional materials.

The new possibilities offered by IN1-Lagrange in the low energy range are illustrated in Fig. 5 by measurements with the Si111 face of the monochromator. The two transitions between energy levels of $Pr^{3+}$ ion split by the electric crystal field are clearly seen in the spectrum recorded with a resolution of ~0.8 meV limited in this energy range mainly by the analyser. We estimate that with IN1-Lagrange useful information on the excitation spectra can be obtained starting from energy transfers as low as ~1.5 meV.

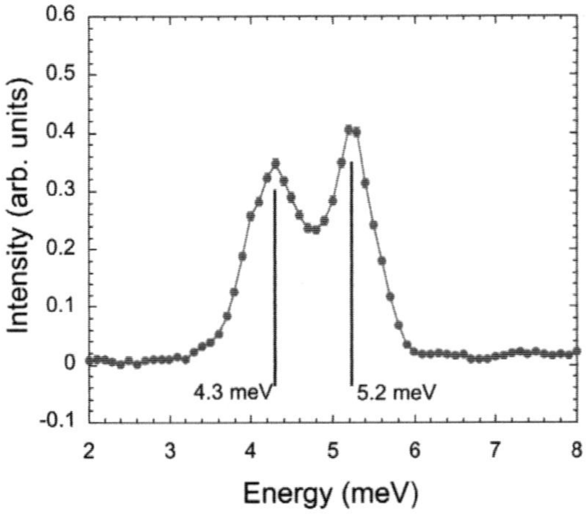


**Figure 5.** The spectrum of low-energy crystal field excitations of $Pr^{3+}$ ions in a test sample taken at T=4 K with Si111 monochromator reflection. The horizontal axis represents the neutron energy loss in the scattering.

The luminosity power of IN1-Lagrange has been challenged with a small sample of 2,5-diiodothiophene. The energy spectrum of the sample containing only 100 microgram of hydrogen is presented in Fig. 6. All the principal vibration lines can be identified in spite of some noise present between the lines. Such "vanishingly small" quantities of scattering matter have so far been considered as impossible for neutron spectroscopy in this energy range. The excellent performance of IN1-Lagrange with small samples is confirmed in subsequent experiments (*e.g.* [9], and others).

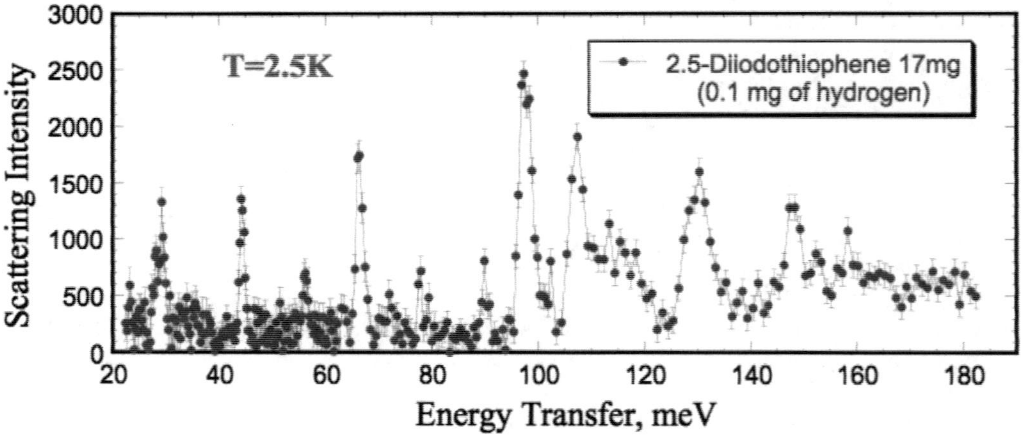


**Figure 6.** The spectrum of a 17 mg 2,5-diiodothiophene sample (containing 0.1 mg of hydrogen) obtained as a difference of the scattering intensities measured on IN1-Lagrange with the sample and with an empty sample can. Both runs in the shown energy transfer range took less than 8 hours each with the Cu220 2D-focussing monochromator.

We outline that with IN1-Lagrange, in contrast to the time-of-flight spectrometers, one can concentrate on a selected energy interval when detailed knowledge of the entire spectrum is not required. Thus the available beam time can be used for improving statistics in the most important spectral region of the measured VDOS or for detailed records of the effect of the sample environment conditions (variation of temperature, pressure *etc*.).

It is clear that the possibility of measuring good quality spectra from very small samples, including tempting studies at high pressures, will be strongly challenged by the instrument background. Our preliminary data are promising and we expect further progress in characterisation and reduction of this important component of the instrument performance for research with extremely small samples.

## 4. Conclusion

The new spectrometer IN1-Lagrange at the hot neutron source of the ILL high flux reactor has been successfully commissioned and is now in routine use for user experiments. The boost of sensitivity offered by IN1-Lagrange opens new possibilities in spectroscopic studies of modern functional materials. The new instrument permits investigation of much smaller sample quantities, inherent in novel materials studies, and a more detailed probing of external parameters like temperature, pressure, chemical composition, time-dependent processes *etc*. The improved energy resolution provides access to a higher level of detail in the vibrational spectra over the whole energy range, thus stimulating further development of theoretical models and *ab-initio* calculations.

## References

[1] Windsor C G 1986 Experimental techniques *Methods of Experimental Physics* (vol 23A: Neutron Scattering) ed K Scold and D L Price (Academic Press) chapter 3 pp 197-257

Arai M 2013 Experimental techniques *Experimental Methods in the Physical Sciences* (vol 44: Neutron Scattering - Fundamentals) ed F Fernandez-Alonso and D L Price (Elsevier) chapter 3 pp 245-320

[2] http://www.isis.stfc.ac.uk/instruments/tosca/

[3] http://flnp.jinr.ru/134/; http://flnp.jinr.ru/391/; http://flnp.jinr.ru/384/

[4] http://neutrons.ornl.gov/vision/ and http://neutrons.ornl.gov/vision/publications/

[5] http://lansce.lanl.gov/lujan/instruments/FDS.shtml

[6] http://www.ncnr.nist.gov/instruments/fans/

[7] http://www.ill.eu/

[8] Rossi Albertini V, Colognesi D and Tomkinson J 2000 *J. Neutron Research* **8** 245

[9] Mukhopadhyay S, Gutmann M J, Jura M, Jochym D B, Jiménez-Ruiz M, Sturniolo S, Refson K and Fernandez-Alonso F 2013 *Chem. Phys.* **427** 95

Horsewill A *et al* 2014 *private communication; to be published*

# Parameters of the NERA spectrometer for cold and thermal moderators of the IBR-2 pulsed reactor

**I Natkaniec[1,2], D Chudoba[1,2], Ł Hetmańczyk[1,3], V Yu Kazimirov[1], J Krawczyk[1,4], I L Sashin[1] and S Zalewski[1,5]**

[1]Joint Institute for Nuclear Research, Frank Laboratory of Neutron Physics, Joliot-Curie 6, 141980 Dubna, Russian Federation
[2]Adam Mickiewicz University, Faculty of Physics, Umultowska 85, 61-614 Poznań, Poland
[3]Jagiellonian University, Faculty of Chemistry, Ingardena 3, 30-060 Kraków, Poland
[4]H. Niewodniczański Institute of Nuclear Physics, Polish Academy of Sciences, Radzikowskiego 152, 31-342 Kraków, Poland
[5]Siedlce University of Natural Sciences and Humanities, Institute of Chemistry, 3 Maja 54, 08-110 Siedlce, Poland

E-mail: inat@jinr.ru

**Abstract**. The inverted geometry time-of-flight spectrometer NERA designed for simultaneous investigation of elastic and inelastic neutron scattering was constructed on a 100 meter long flight path of the IBR-2 reactor at JINR, Dubna, Russia. Spectrometer parameters were optimized for the thermal neutron spectrum of a water moderator at 330 K. Recently, a new type of cold neutron source operating at 30 K was installed in the sector of six horizontal channels. The cold source in combination with the water moderator allows one to effectively use incident neutrons in wide range of wavelengths.

## 1. Introduction

A 100 meter long flight path on channel no. 7 at the IBR-2 pulsed reactor was constructed in the 1980's for high resolution studies of condensed matter by neutron powder diffraction (NPD), quasielastic (QENS) and inelastic neutron scattering (INS). Two home-made direct Ni-mirror guides with a transversal cross section of the $5 \times 16$ cm$^2$, extracted neutron beams 7a and 7b, for instruments located at 103 m and 109 m flight path, respectively. Mirror guides were installed at a distance of 7.5 m from the IBR-2 water comb-like moderator. Beams 7a and 7b crossed at the window of the fast neutrons background chopper located at 5.5 m from the moderator. The background chopper at such geometry of extracted beams was used also as a neutron wavelength selector ($\lambda$-chopper) that transmitted the same thermal spectrum for both beams. At 100 m of flight path and a frequency of 5 Hz of IBR-2 pulses, the neutron spectrum must be limited to 7.5 Å to avoid frame overlap. The beam 7a was used for the development of new instruments for effective investigations of crystalline texture and internal stress analyses by the time-of-flight high resolution neutron powder diffraction method. Finally, two modern instruments: SCAT – for texture analysis and EPSILON – for internal stress analysis were constructed [1]. The NERA-PR spectrometer [1,2], located on the beam 7b at a distance of 109.05 m from the moderator was commissioned in 1991 and operated effectively until the end of 2005, when the IBR-2 reactor was shut down for modernization.

A new type of cold neutron source for the sector of channels 7-11 was constructed during the modernization period 2006-2011 [3]. A cold moderator medium, an organic glass of mesitylene and m-xylene solution was selected as a result of investigations of the structural properties and methyl group dynamics of solid phases of methyl derivatives of benzene [4, 5], which were performed on the NERA-PR spectrometer. The modernization program for spectrometers located on 100 m flight paths of channel no. 7 included the replacement of the old neutron guides and extraction of three independent beams. The three new mirror guides begin after the background chopper window at a distance of 6 m from the moderator. The first 10 m long common vacuum section which contains three mirror guides with direct visibility of moderator is called the 'splitter'. Independent neutron guides (7a1 - EPSILON, 7a2 – SCAT and 7b – NERA) begin at a distance of 16.5 m from the moderator.

In this paper we compare the parameters of the NERA spectrometer before and after the modernization of the IBR-2 pulsed reactor, including the new cold moderator and the replacement of the neutron guides.

## 2. Modernization of the NERA spectrometer

The layout of the inverted geometry NERA spectrometer is shown in figure 1. The primary spectrometer defines the energies of the incident neutrons and transports the neutron beam from the IBR-2 moderator (2) to the sample position (11) at $L_1$ = 109.05 m. A new fast neutron background chopper (4) rotates around a horizontal axis and is located in the inner corridor of the IBR-2 reactor shielding at an unchanged distance of 5.5 m from the moderator. The chopper disc has the same diameter of 134 cm, however, the thickness of the $TiH_2$ layer which removes fast neutrons was doubled to 6.2 cm. The rotating disc window forms the $60^{\circ}$ sector free of $TiH_2$ layer. The present geometry of the extracted beams, which are separated at the background chopper window, makes it impossible to use the λ-chopper mode of the background chopper satisfactorily for all three spectrometers due to different phases.

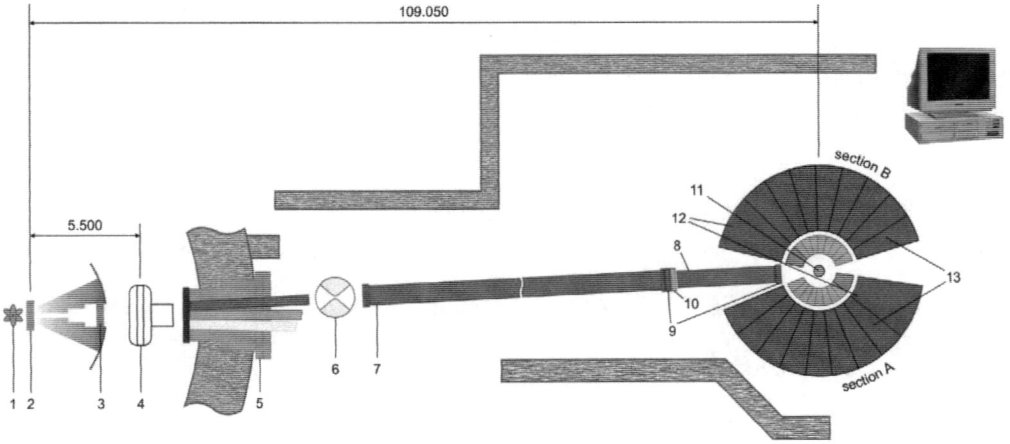


**Figure 1.** The layout of the NERA spectrometer: 1 – IBR-2 reactor core, 2 – thermal and cold moderators of radial horizontal channels 7-11 and tangential channels 1–9, 3 – beam shutter, 4 – fast neutron background chopper, 5 – common vacuum splitter of three Ni-mirrors neutron guides, 6 – λ-chopper of beam 7b, 7 – vacuum Ni–mirrors guide tube of neutron beam 7b, 8 – vacuum sections of beam 7b, 9 - diaphragms of incident beam, 10 –monitor, 11 – sample position, 12 – NPD sections, 13 – INS and QENS sections.

The splitter (5) transports three neutron beams from the inner corridor to a distance of 16.5 m out of the biological shielding of the reactor. From this place, the neutron beams are transported by independent vacuum Ni-mirror guide tubes (7).

Due to the fact that experiments with cold neutrons up to $\lambda$ = 14 Å were planned on beams 7a1 and 7a2, a new velocity selector (λ-chopper) was constructed and installed at a flight path distance of $L_\lambda$ = 26.95 m from moderator. The λ-chopper drum (diameter = 48 cm) rotates in the vertical axis at a frequency of 2.5 Hz, which is half that of the frequency of the IBR-2 pulses. A TOF diagram, transmission function, as well as a general view and rotor window shape are presented in more detail in figure 2.

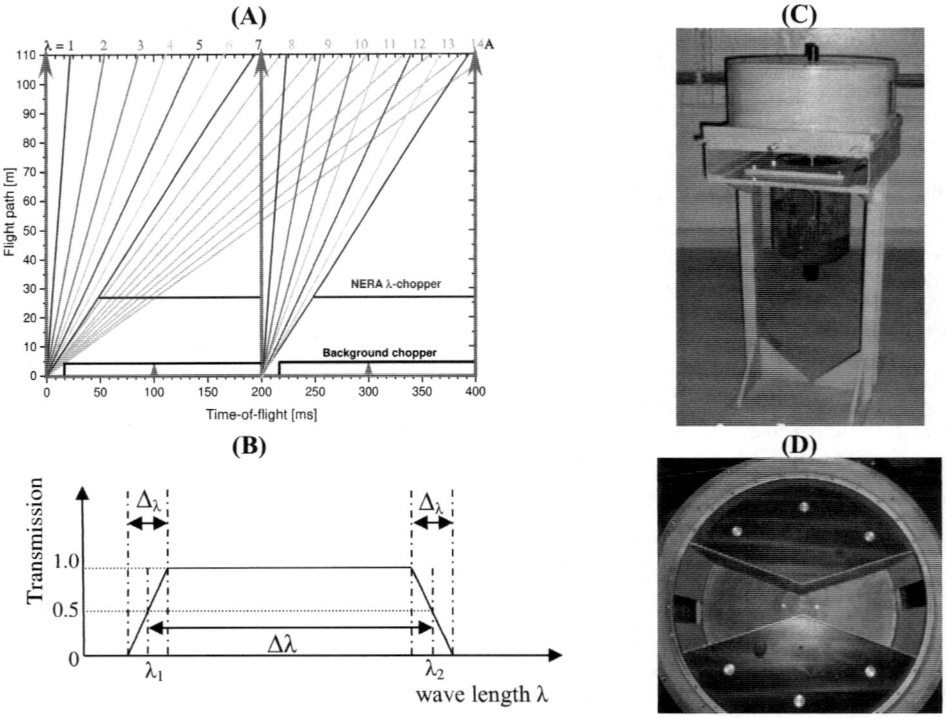


**Figure 2**. TOF diagram - (A) and transmission function - (B) of the λ-chopper beam 7b, where $\lambda_2 - \lambda_1 = \Delta\lambda = 7.1$Å, $\Delta_\lambda = 0.975$Å. General view (C) and the drum rotor (D) of installed λ-chopper.

As part of the modernization, the evacuated Ni-mirror guide tube was replaced. The transversal dimension of the new neutron guides is 5 × 16 cm$^2$ as it was before. The main part of the 80 m long vacuum guide tube ends 2 m before the sample position. An additional vacuum section 1.5 m long includes a monitor counter and a beam-shaping diaphragm in the incident beam.

The design features of the main part of the NERA spectrometer, which analyses and records the scattered neutrons, are presented in figure 3. The spectrometer consists of two symmetrical sections, A and B. One scattering chamber for NPD and eight chambers for INS and QENS measurements are located in each of them. For NPD measurements neutrons scattered by the sample (1) pass through collimators (3) and are detected with two kinds of detectors: cylindrical He-3 counters (7) without additional collimation, which record NPD spectra at high luminosity, and rectangular He-3 counters (8) positioned after long Soller collimators, which record NPD spectra at a low background level. The NPD spectra are recorded at scattering angles of $30^{\circ} \leq 2\theta \leq 70^{\circ}$ and $110^{\circ} \leq 2\theta \leq 150^{\circ}$.

Details of the INS and QENS scattering chambers are presented in figure 3. The neutrons scattered by a sample (1) pass through the common beryllium filter collimators (2), which are cooled by liquid nitrogen. At this temperature beryllium only transmits the neutrons with energy less than 5.25 meV. The INS and QENS spectra are recorded at scattering angles from 20° to 160° in 10° increments.

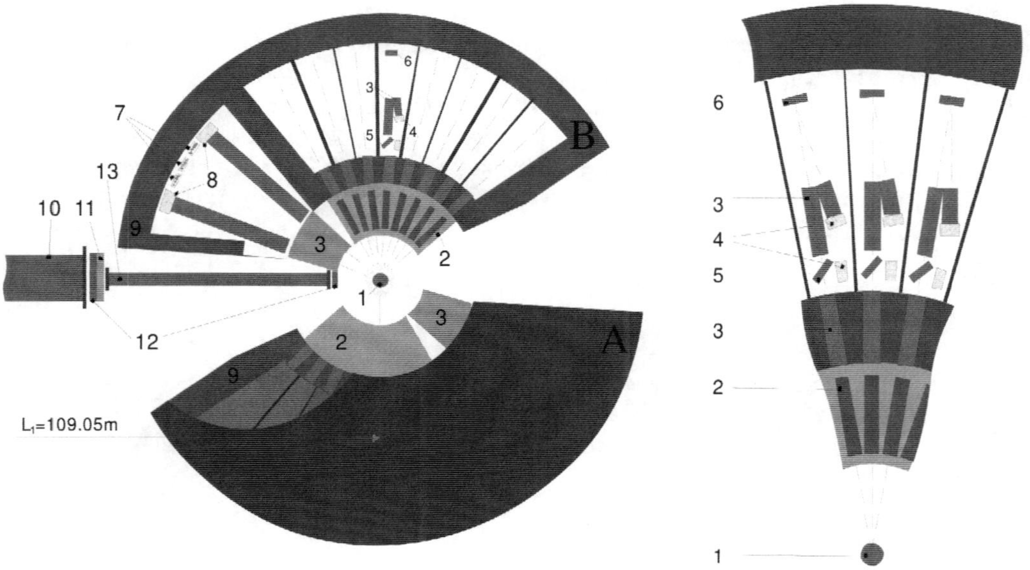


**Figure 3.** The main part of the NERA spectrometer: 1 - sample, 2 - Be-filters, 3 - collimators, 4 – He-3 detectors (INS and QENS), 5 - PG analysers (INS), 6 - single crystal analysers (QENS), 7 – detectors for high intensity diffraction, 8 – diffraction detectors with a good collimation, 9 - spectrometer shielding, 10 - Ni-coated mirror neutron guide in a vacuum tube, 11 - incident beam monitor, 12 – diaphragms, 13 - vacuum neutron guide.

There are three possible ways to analyse the energy of scattered neutrons with different resolution:
- by using the beryllium filters and detectors located directly behind them. In this case the spectrometer has high luminosity at low energy resolution,
- by using beryllium filters and pyrolytic graphite – PG(002) analysers at the scattering angle $2\Theta = 90^{\circ}$ before the detectors. This is a main option of spectrometer for measurements of the INS spectra with medium resolution and reasonable intensity,
- by using beryllium filters and crystal analysers (Cu, Al, Zn or PG) at the scattering angle $2\Theta = 174^{\circ}$. In this case the INS luminosity is very limited but the width of elastic lines (40 - 60 μeV) is suitable for investigations of quasielastic neutron scattering. This mode can be carried out simultaneously with the second one, because the elastic line positions of the crystal analysers at backscattering differ from those of the neutron energies reflected by pyrolytic graphite at $2\Theta = 90^{\circ}$.

The modernization of the NERA automatic control instruments and TOF analyser electronics, as well as operational and computing software for recording, processing and visualization of experimental data will be the subject of a separate paper.

## 3. Time of flight incident neutron spectra

A low efficiency gaseous proportional counter filled with a mixture of nitrogen and inert gases is used for monitoring the incident neutron flux on the NERA spectrometer. The time-of-flight (TOF) spectra of incident neutrons measured by the monitor counter are shown in figure 4.

Spectrum A was measured with fully open windows of stationary background and λ-choppers. One can see that it was recorded at a rather high background of overlapping neutrons, which have lost correlation with the reactor pulses and diffused inside the mirror guide. This background can be effectively removed by proper phasing of λ-chopper with the IBR-2 pulses, as it is shown by spectra B and C. At phase F1, which corresponds to spectrum B, the transmission window of λ-chopper shown in figure 2, is fully open at the time of the IBR-2 pulse. The bent mirror guide does not transfer fast and epithermal neutrons up to 16 ms after the reactor pulse, which corresponds to the neutron wavelength of $\lambda$ = 0.6 Å. Spectrum C, recorded with phase F2 = F1 + 16 ms, corresponds to the optimal transfer without overlapping of the wavelength range $\lambda$ = (0.6 - 6.6) Å, as was intended in the design of the λ-chopper.

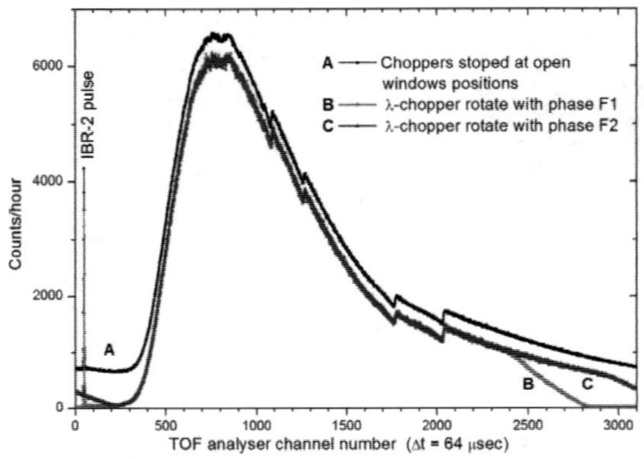


**Figure 4.** The TOF incident neutrons spectra measured by monitor counter for water moderator (WM) at different phases of the λ-chopper.

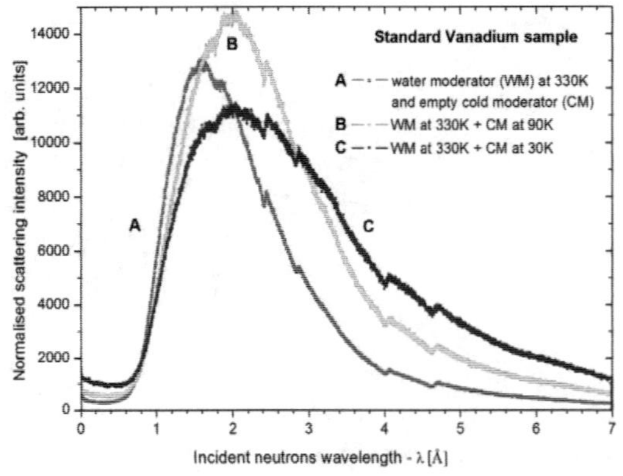


**Figure 5.** The TOF spectra measured by diffraction detectors after scattering by vanadium sample for water (WM) and cold moderator (CM) at different temperatures.

Incident neutron spectra of the water and cold moderators as measured by the diffraction detectors after scattering from the standard vanadium sample are shown in figure 5. The standard sample was prepared from 100 g of vanadium foil, rolled in a shape of a cylindrical tube 5 cm in diameter and 20 cm high. This sample matches the transversal cross-section of the neutron guide and scatters all incident neutrons at the sample position. Using this standard vanadium sample we can determine the efficiency of the detector system and compare parameters of the NERA spectrometer for a long time period.

The TOF incident neutron spectra measured before and after the reconstruction of the IBR-2 are compared in figure 6 in a wavelength scale. The TOF scale has been transformed into the incident neutron wavelength by using the following relation:

$$\lambda = h/mv \quad \text{or} \quad \lambda[\text{Å}] = 3.956 \times 10^{-3}\, t[\mu\text{s}]/L[\text{m}],$$

where: $h$ is Planck's constant, $\lambda$, $m$ and $v$ - are the neutron wavelength, mass and velocity respetively, $t$ - is the time of flight and $L$ – is the average total distance from the IBR-2 moderator to the NERA diffraction detectors. The average total distance $L$ can be calculated from the cut-off edges of the aluminium windows in the path of the neutrons from the moderator to detectors, which are clearly seen in all incident neutron spectra. These edges correspond to the neutron wavelength

$\lambda$(hkl) = 2$d$(hkl) of the corresponding interplanar distances of the Al crystal structure. The calculated average value of $L$ is (110.47 ± 0.1) m. The NPD detectors (He-3 counters, 2 cm thick) are placed at the radius $L_{2D}$ = (1.415 ± 0.01) m from the centre of the sample position. The flight path of the NERA spectrometer calculated in this way has a value equal to $L_1 = L - L_{2D}$ = (109.05 ± 0.05) m.

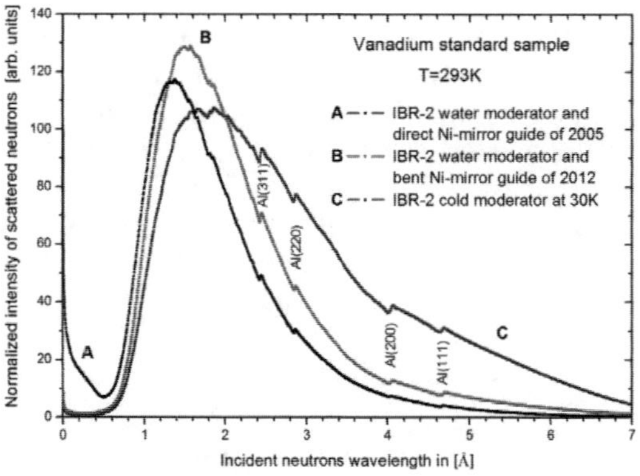


**Figure 6.** Incident neutrons spectra before and after modernization of the IBR-2 moderator and the Ni-mirror vacuum neutron guides of the beam 7b, normalized to average power of the reactor.

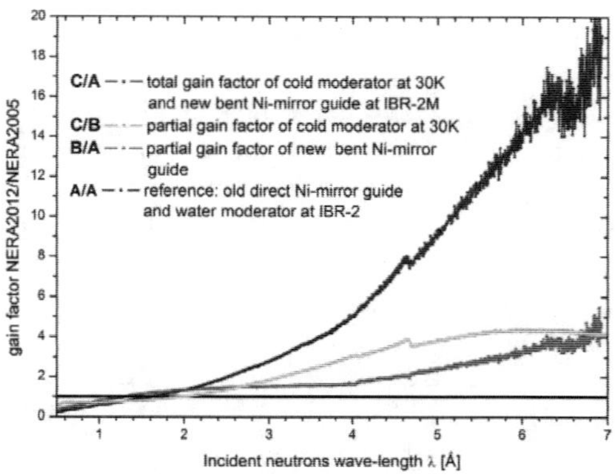


**Figure 7.** Gain factors versus incident neutron wavelength after the modernization of the IBR-2 moderator and the Ni-mirror neutron guides of the NERA spectrometer.

By comparing the spectra shown in figure 6, one can conclude that the mirror guide 7b has no direct visibility of the IBR-2 moderator, since the epithermal neutrons present in spectrum A are absent in spectra B and C. A direct neutron guide was planned in the modernization project of beam 7b, however, a mistake was made when the splitter was adjusted in optimal way for the neutron beams 7a1 and 7a2. As a result it was not possible to conduct the direct beam 7b to the experimental pavilion and the last 80 m section of the NERA mirror guide had to be slightly bent in order to transport neutrons to the sample position.

The integrated values of spectral distributions shown figure 5 in the range of $\lambda$ = (0.5 - 6.5) Å have the following ratio relations: B/A = 1.34 and C/A = 1.27. The reference spectrum A corresponds to the water moderator and an empty cold moderator container. Spectra B and C correspond to the fully loaded container of the cold moderator with mesitylene pellets at 90 K and 30 K, respectively. One can see that the integrated neutron flux increases by 34% when the cold moderator container was filled with mesitylene at 90 K. Further cooling of moderator to 30 K, slightly decreases the integrated gain factor to 1.27, because cold neutrons with $\lambda$ > 6.5 Å are beyond the integration limits.

The integrated values of the spectra shown in figure 6 are in the ratio: B/A = 1.2 and C/A = 1.6. Here the reference spectrum A corresponds to the comb-like water moderator in 2005, the year before the reconstruction of the IBR-2 reactor core. The B and C spectra correspond to the new thermal water moderator and the cold moderator at 30K, respectively. The value 1.2 of the integrated gain factor of the water moderator thermal neutron flux means that the new bent mirror guide transports 20% more neutrons than the old direct mirror guide. The better quality of the new Ni-mirror neutron guide also contributes to the fact that the neutron flux at the sample position from the cold moderator is greater by 60% as compared to the old comb-like water moderator.

The partial and total spectral gain factors versus incident neutron wavelength are shown in figure 7. From the comparison with the reference spectrum of neutrons transported by the old direct mirror guide in 2005, one can see that the partial gain factor of the new bent mirror guide for water moderator becomes grater then one for $\lambda$ > 1.3 Å, increases twice at $\lambda$ = 4 Å, and further increases linearly to five at λ = 7 Å. The partial gain factor of the new cold moderator over the new water moderator becomes greater than one at λ = 2 Å, linearly increases to 4 at λ = 4.7 Å and remains at this value up to λ = 7 Å. The total gain factor of the new cold moderator and Ni-mirror guide in 2012 compared to the old moderator and mirror guide in 2005 becomes greater than one at λ = 1.5 Å and increases up to about 20 at λ = 7 Å.

## 4. Energy analysis of the scattered neutrons and resolution of the NERA spectrometer

A set of energy analysers of the scattered neutrons consists of polycrystalline beryllium filters cooled to liquid nitrogen temperature and pyrolytic graphite (PG) or single crystal Cu or Zn in front of the detectors. Be-filters eliminate higher order reflections of the basal plane of crystal analysers. The Be-filters can be used as the energy analysers for a high intensity INS mode at low resolution of the spectrometer when the detectors are placed directly after them. Spectral distributions of neutrons scattered by a standard vanadium sample and registered by the helium-3 detectors are shown in figure 8, versus incident neutrons wavelength, and in figure 9, versus final energy scattered neutrons.

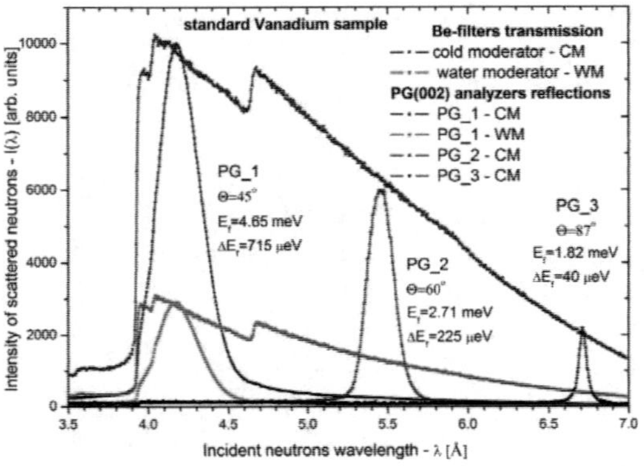


**Figure 8.** Transmission of the Be-filters and reflections of the PG(002) analyzers at different Bragg angles - Θ.

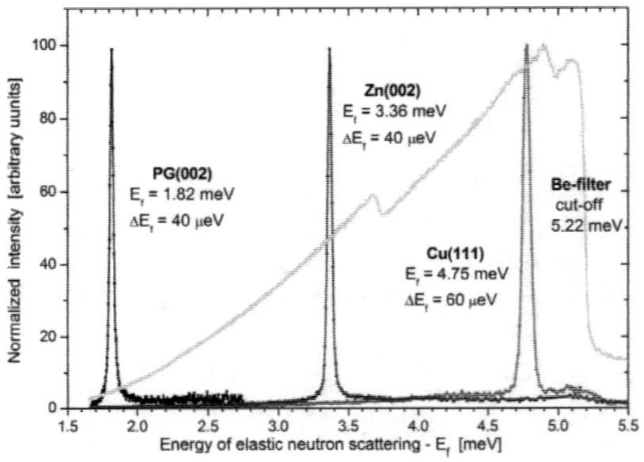


**Figure 9.** Final energies and FWHM of elastic peaks of different crystal analyzers at nearly back scattering reflection - 2Θ = 174°.

The significant gain of neutron flux in the wavelength range of $\lambda$ = (4 - 7) Å given by the cold moderator, as shown in figure 7, enables us to use crystal analysers at nearly back scattering for INS and QENS spectra in the high resolution mode of the NERA spectrometer (see figure 9).

The intensity of neutrons $I(t,\varphi,T)$ which were scattered at an angle $\varphi$ and sample temperature $T$, and recorded at time $t$ can be expressed as a convolution of the scattering cross section $\sigma(E_i,E_f,\varphi,T)$ with the incident neutron flux $\Phi(E_i)$ and the instrumental resolution $R(E_i, E_f, t_0, t)$

$$\boldsymbol{I(t,\varphi,T)\Delta t = \Delta t \iiint \sigma\left(E_i,E_f,\varphi,T\right)\Phi(E_i)R\left(E_i,E_f,t_0,t\right)dE_i dE_f dt_0} \quad (1)$$

where: $\Delta t$ is the time interval corresponding to the channel width of the time-of-flight analyser, $E_i$ and $E_f$ are incoming and scattered neutron energies, respectively.

The resolution function of NERA spectrometer can be written as:

$$\boldsymbol{R\left(E_i,E_f,t_0,t\right) = \rho(E_i,t_0)n\left(E_f\right)\delta(t-t_0-t_1-t_2)} \quad (2)$$

$\rho(E_i, t_0)$ is the time distribution of incoming neutrons with the energy $E_i$ leaving the source at time $t_0$. It can be calculated as a convolution of the fast neutron pulse shape with a function that describes the moderation process of neutrons. The fast neutron pulse shape of the IBR-2 is a Gaussian function with FWHM equal to 215 μs, and the parameters of moderation function for the comb-like water moderator were determined in [6].

$n(E_f)$ is the energy distribution of scattered neutrons, as shown in figures 8 and 9.

$\delta(t-t_0-t_1-t_2)$ gives the condition for registration of neutrons at time $t$, which left the source at time $t_0$ with energy $E_i$ and were scattered to energy $E_f$.
$t_1=\alpha L_1/\sqrt{E_i}$ is the flight time of neutron with energy $E_i$ of the flight path $L_1$ = 109.05 m.
$t_2=\alpha L_2/\sqrt{E_f}$ is the flight time of the scattered neutron with energy $E_f$ from the sample to detector distance $L_2$ = 1.015m,
$\alpha$ = 2286.27 is a conversion coefficient of neutron energy to velocity when time is expressed in [μs], flight path in [m] and energy in [meV].
The resolution functions of the different components of inverted geometry spectrometers are presented in more detail in [7].

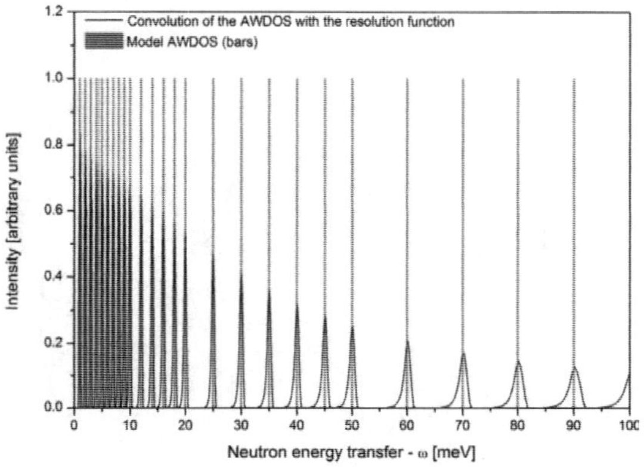


**Figure 10.** Shape of resolution functions of the NERA spectrometer for PG(002) analyzers reflection at scattering angle 2Θ = 120°, calculated according to eq. (1).

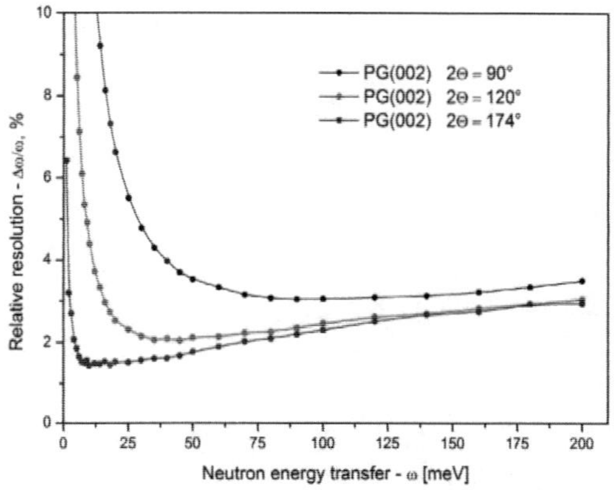


**Figure 11.** Relative resolution of the INS spectra measured with PG(002) analyzers at different scattering angles 2Θ = 90°, 120° and 174°.

The scattering cross-section for phonon creation process at low temperatures can be written in the incoherent one-phonon scattering approximation in the following way

$$\sigma_1^{inc}(E_i, E_f, \varphi, T) \approx \sqrt{\frac{E_f}{E_i}} \frac{\hbar |Q(E_i, E_f, \varphi)|^2}{\omega} \sum_n \frac{(b_n^{inc})^2}{M_n} \frac{exp(-2W_n)}{1-exp\left(-\frac{\hbar\omega}{k_B T}\right)} G(\omega) \quad (3)$$

where: $Q(E_i, E_f, \varphi)$ is the neutron momentum transfer, $\omega=(E_i-E_f)$ is the neutron energy transfer, $b_n^{inc}$ is the value of the incoherent scattering length, $M_n$ is the mass of the $n^{th}$ atom, $exp(-2W_n)$ is the Debye-Waller factor and $G(\omega)$ is the so-called amplitude weighted phonon or vibrational density of state (AWDOS), defined as:

$$G(\omega) = \sum_j \int d^3q \left[A_j^n(q,\omega)\right]^2 \delta\left[\omega - \omega_j(q)\right] \quad or \quad G(\omega) = \sum_j \sum_n \left[A_j^n(\omega)\right]^2 \delta\left(\omega - \omega_j\right) \quad (4)$$

$A_j^n(\omega)$ is the amplitude of displacement of the $n^{th}$ atom in the unit cell at the vibration mode $\omega_j(q)$, $j$- is an index that runs over all the dispersion curves or internal normal modes and **q** is a reciprocal lattice wave vector in the Brillouin zone.

Computational models, which give frequencies $\omega_j(q)$ and displacement amplitudes $A_j^n(q,\omega)$ of atomic vibrations in the crystal unit cell or in atomic or molecular clusters, allow one to calculate directly the INS TOF spectra in the approximation of one-phonon incoherent scattering process, according to equations (1), (2), (3) and (4).

Assuming the model of the AWDOS functions, in the form of δ-functions at a given energy transfer is as shown in figure 10, one can directly calculate the shape of the resolution function of the NERA spectrometer according to equation (1), by convolution of experimentally determined energy distribution functions: $\Phi(E_i)$, $\rho(E_i, t_0)$ and $n(E_f)$ at the TOF condition $\delta(t-t_0-t_1-t_2)$. The relative resolution functions of the INS spectra $\Delta\omega/\omega$ versus neutron energy transfer $\omega=E_i-E_f$ are presented in figure 11.

## 5. Background ratio and luminosity of inelastic scattering and neutron powder diffraction

A standard sample of 10 ml of water in a rectangular 1 × 60 × 160 mm$^3$ Al sample holder was measured with thermal and cold moderators in order to check the correctness of the transformation of TOF INS spectra into the AWDOS functions G(ω). The background level of an empty sample holder and the cryostat in this case do not exceed several percent of the INS intensity. However, the water sample produces a very high incoherent background at the NPD detectors. The TOF INS spectra of ice, measured at 20 K with PG_1 analysers (2Θ = 90°) and summed over 15 scattering angles in the range from 20° to 160°, are compared in figure 12.

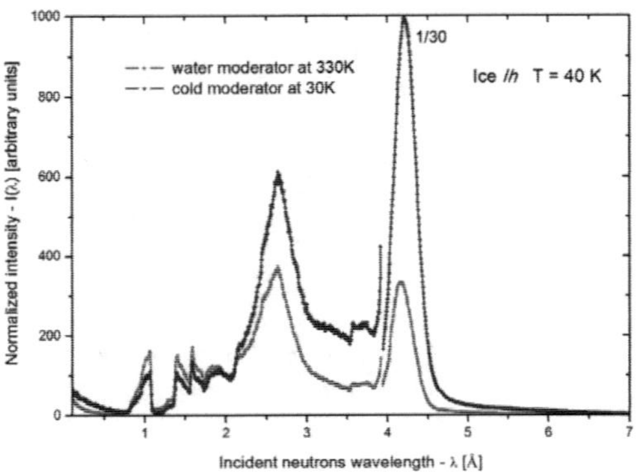


**Figure 12.** The experimental TOF INS spectra of Ice *Ih* measured at T=40 K for thermal and cold moderators of the IBR-2 pulsed reactor.

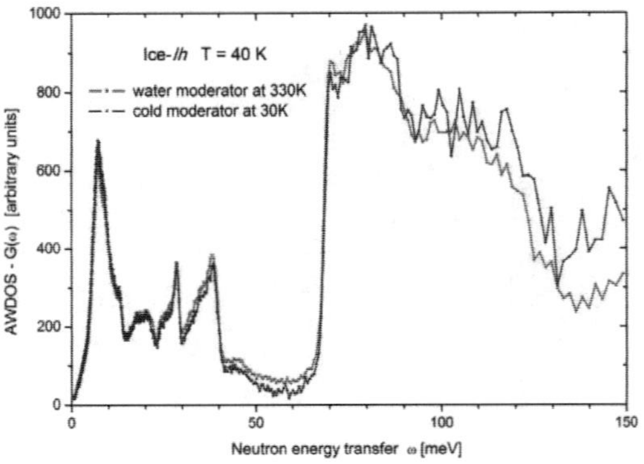


**Figure 13.** Amplitude-weighted density of phonon states (AWDOS) $G(\omega)$ of Ice *Ih* obtained from the TOF INS spectra from figure 12.

The spectra measured with thermal and cold moderators were normalized to 10 hours measurement time at a constant reactor power and similarly processed background spectra were subtracted. After such normalization we could not obtain consistent $G(\omega)$ functions of ice for both moderators. The reason was that the incident neutron fluxes of both moderators were normalized to unity at their maxima by the NERA software, while they have the same values only at the wavelength $\lambda = 2$ Å (see figure 6). The TOF INS spectra normalized to integrated monitor counts or measurement time do not fulfil this condition for normalization, as well. Additional normalization of the TOF INS spectra of ice to the same value at $\lambda = 2$ Å results in almost identical $G(\omega)$ functions shown in figure 13.

The transformation of the TOF INS spectra to $G(\omega)$ is done omitting the convolution procedure and using average values of spectral distribution functions written in equations (1) and (2). The IBR-2 pulse shape under this assumption is characterized by $\delta(t - T_0)$, where $T_0$ is an average delay time between the electronic pulse starting the TOF analyser and the real neutron pulse of the reactor. The average energy of incident neutrons $E_i(i)$, where $i$ is the channel number of TOF analyser, is calculated by the relation introduced in section 4:

$E(i)$ [meV] = $(\alpha L_1[\text{m}]/t_1(i)[\mu\text{sec}])^2$, where: $t_1(i) = (i - 1/2)\Delta t - T_0 - t_2$.

The final energy of the scattered neutrons $E_f$ depends on the particular crystal analyser used; their spectral distributions and average values are shown in figures 8 and 9.

The intensity of the scattered neutrons in the TOF INS spectra, $I(\lambda)$, normalized to incident neutron spectra, $\Phi(\lambda)$, defines the scattering cross-section. For hydrogen-containing materials the incoherent inelastic scattering of hydrogen atoms dominates the scattering power of other atoms. By assuming one-phonon incoherent scattering process and using average values of *E(i)* and $E_f$ one can transform the INS TOF spectra to the generalized density of vibrational states function $G(\omega)$ directly by using equation (3).

The experimental conditions for investigation of materials with low coherent scattering cross-sections can be illustrated by recently measured INS and NPD spectra of graphitized mesoporous carbons (GMCs) and graphene oxide (GO), which are compared with the reference spectra of spectroscopic purity (99.99%) graphite powder (Grs) in figures 14 and 15.

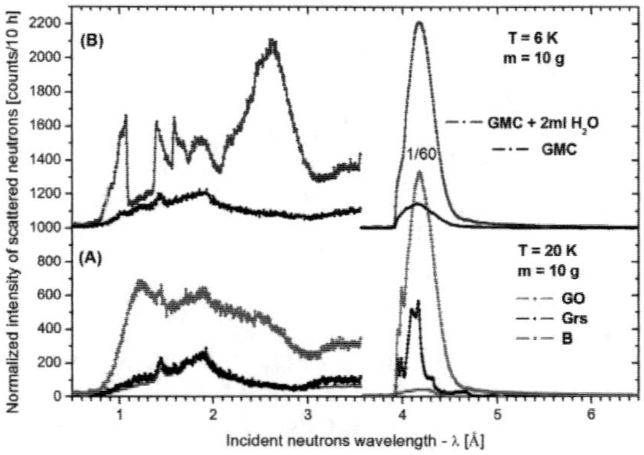


**Figure 14.** The INS TOF spectra: (A) - empty cryostat – B, spectral graphite – Grs and GO; (B) - GMC and GMC with pores filled by water.

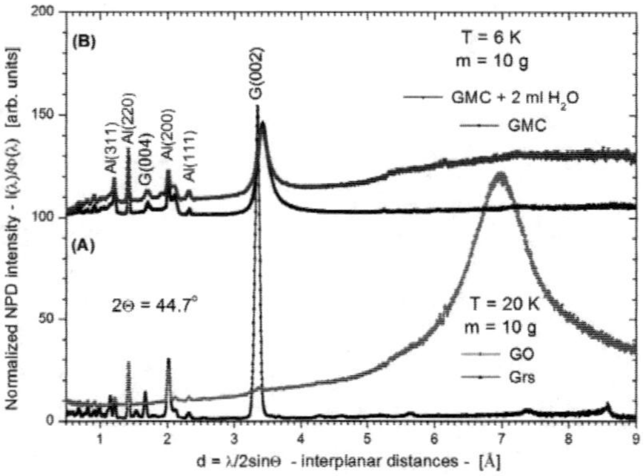


**Figure 15.** Neutron diffraction patterns: (A) - spectral graphite – Grs and GO, (B) - GMC and GMC with pores filled by water.

GMCs just like spectral graphite (Grs) are almost pure carbon hydrophobic materials, so the 10 gram samples are practically invisible in the INS spectra above the background spectrum (B) of the aluminum cryostat and an empty sample holder. The investigated sample of GMC (Aldrich Product No. 699624) is characterized by an average pore diameter of 14 nm and total pore volume of 0.24 $cm^3$ $gram^{-1}$. By adding 2 ml of water to 10 grams of GMC in a hermetically sealed container and warming it for several days at 350 K, a powdered sample of GMC with the pores filled with water was obtained. The INS spectra of dry and wet GMC samples are presented in figure 14. As one can expect from the value of average diameter of pores of investigated GMC material, water frozen in the pores of GMS has similar INS spectrum to that of ice *1h* presented in figure 12.

The chemical composition of GO is close to C/O = 2. The GO contains oxygen and hydroxyl functional groups bound to the basal plane of graphene and is a strongly hydrophilic material [8]. The scattering intensity of the INS spectrum of 10 gram GO sample exceeds by several times the background level. By comparing the integral intensity of the INS spectra of GO and GMC filled with water, one can evaluate the hydrogen atom content in the GO sample at the level of about 2 ml of water. The INS spectrum of GO differs significantly from that of ice *1h* and it is the subject of recent studies by INS and computational quantum chemistry DFT methods [9-11].

The NPD spectra shown in figure 15 were recorded for the scattering angle $2\Theta = 44.7^{\circ}$ at which interplanar distances can be measured up to 9 Å. The recorded spectra $I(\lambda)$ were normalized to integral counts of the monitor and to the spectral distribution of the incident neutrons $\Phi(\lambda)$. Spectral graphite – Grs along with the GMC samples are clearly seen in the NPD spectra because of the high reflectivity of the (002) graphite plane. The diffraction peaks of Al and Grs correspond to resolution function. Much broader peaks observed for the GMC and GO samples reflects the thickness of graphite or graphene stacks in the investigated nano-carbon materials. A broad diffraction peak at about 7 Å in the GO sample spectrum indicates that the interplanar distance between graphene sheets in graphite oxide is twice as large as in graphite. This allows water molecules to migrate between of the graphene oxide sheets.

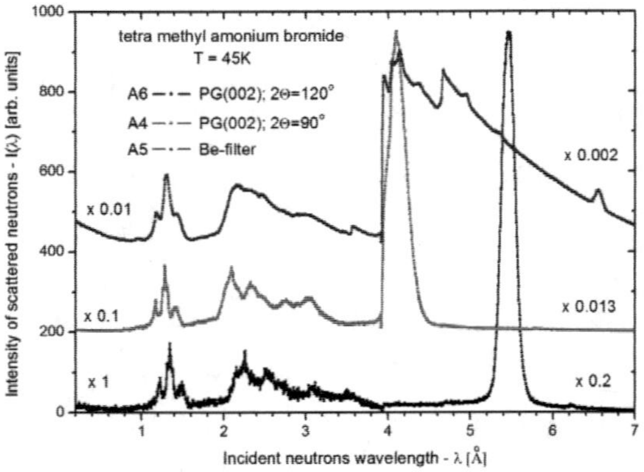


**Figure 16.** Intensity and resolution of INS TOF spectra measured with different analyzers of neighboring detector chambers at scattering angles 130° (A4), 120° (A5) and 110° (A6).

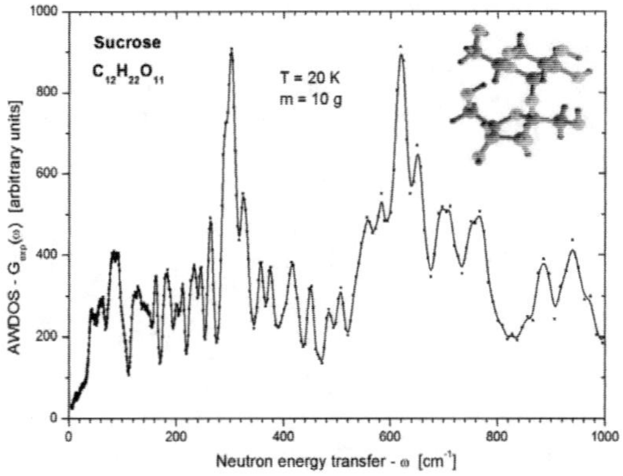


**Figure 17.** Experimental AWDOS $G_{exp}(\omega)$ function of sucrose chosen for illustration of resolution and luminosity of the NERA in the useful frequency range.

The relationship between the resolution and luminosity of the NERA spectrometer are illustrated in figure 16, by the INS spectra of tetramethyl ammonium bromide measured at 45 K. The scattered intensity of the INS TOF spectra measured at neighboring detector chambers for different energy analysers were normalized to 10 hours measurement time. Taking into consideration the energy window of the analysers used, characterized by the FWHM of final energy distributions: A5 – (Be-filter, $\Delta E_f = 2.5$ meV), A4 – (PG at $\Theta = 45°$, $\Delta E_f = 0.715$ meV) and A6 – (PG at $\Theta = 60°$, $\Delta E_f = 0.225$ meV), one can see that the scattered intensity decreases by one order of magnitude when the final energy window narrows by a factor of about three, it means according to the general rule that the luminosity is proportional to the square of the energy window. However, 10 hours measurement time for 10 grams hydrogenous substances seems to be sufficient to record the total spectrum summed over 15 scattering chambers within the resolution window of 0.2 meV. The experimental AWDOS – $G_{exp}(\omega)$ function obtained for sucrose [12] is shown in figure 17 as an illustration of the recent resolution and luminosity of the NERA spectrometer. The frequency range measured with relatively good intensity and resolution of PG analyzers at $\Theta = 45°$ is limited to 1000 $cm^{-1}$ because the bent neutron guide cuts off neutrons with energies above 150 meV.

The results presented in figures 14-17 were obtained in the context of the user program during the first year of operations of the NERA spectrometer after the modernization.

## 6. Conclusions

The main objectives of the modernization program of the NERA spectrometer proposed in 2006 were realized and after modernization of the IBR-2 core in 2011 the spectrometer was commissioned in September, 2012. The gain factor (2012/2005) for the water moderator regarding the average thermal neutron flux at the sample position is 1.2, which is not high; however, the spectral distribution of incident neutrons changed significantly. The gain factor (2012/2005) for elastic scattering with the cold moderator in the range of wavelength 1.5 Å < λ < 7 Å increases from 1 to 20, which provides much better conditions for measurements of neutron powder diffraction (NPD) and quasielastic neutron scattering (QENS) spectra in a high resolution mode with near back-scattering crystal analyzers. Unfortunately the transmission of the new bent Ni-mirror guide limits the wavelength of thermal neutrons to about $\lambda$ = 0.8 Å; as a result the neutron energy transfer for INS spectra is practicably limited to 120 meV (1000 $cm^{-1}$).

The mechanical design of the main part of the NERA spectrometer is convenient from the viewpoint of flexibility of replacement and adjustment of the crystal analysers in order to adapt their resolution to the experimental needs. The arrangement of the Be-filters and crystal analysers in the horizontal scattering plane significantly limits the luminosity of the NERA analyser. The horizontal scattering plane determined by the central level of incident and scattered beams does not make it possible to increase the effective area of the crystal analysers over the area of the Be-filters windows. The NERA crystal analyser system could be combined with the high luminosity neutron energy analyser LAGRANGE recently constructed at the ILL [13]. Such a combined system would effectively use neutrons scattered in circular geometry both in and down of the horizontal plane. However, such project seems to be too expensive for realization at the IBR-2 in near future.

The simultaneous measurements of the INS, QENS and NPD spectra with different resolutions of the NERA crystal analyzers give a unique possibility for observing the dynamics and structure of investigated samples, what it is very useful for investigations of organic compounds with rich polymorphism under different thermodynamic conditions.

**Acknowledgements**
The authors are sincerely grateful to the late Professor J.A. Janik and to Professors A.V. Belushkin, W. Nawrocik and J. Wąsicki, as well as to Doctors D.P. Kozlenko and S.A. Kulikov for their keen interest and support of the modernization project of the NERA spectrometer. We thankfully acknowledge the efforts of those who have assisted in the construction and adjustments of a new Ni-mirror guide and $\lambda$-chopper of beam 7b including engineers A.P. Bulkin and V.A. Kudryashov from PINP at Gatchina, engineers A.P. Buzdavin, A.N. Chernikov, A.P. Sirotin and V.V. Zhuravlev and their team of technicians from the Neutron Spectrometers Service Department of FLNP JINR who supported the operation and maintenance of cryogenic, electronic, mechanical and vacuum equipment of the NERA spectrometer.
Grants from the Polish Plenipotentiary in JINR for the modernization project of the NERA spectrometer in 2007-2012 are gratefully acknowledged.

**References**
[1] http://flnp.jinr.ru, FLNP User Guide.
[2] Natkaniec I, Bragin S I, Brańkowski J and Mayer J 1994 in *Proc. ICANS XII Meeting*, Abingdon 1993, Rutherford Appeton Laboratory Report 94-025, **Vol. I** 89-95.
[3] Ananiev V, Belyakov A, Bulavin M, Kulagin E, Kulikov S, Mukhin K, Petukhova T, Sirotin A, Shabalin D, Shabalin E, Shirokov V, Verhoglyadov A 2014 *Nuclear Instruments and Methods in Physics Research B* **320** 70-74.
[4] Natkaniec I, Hołderna-Natkaniec K and Kalus J 2004 *Physica B* **350** 651-653.
[5] Natkaniec I, Shabalin E, Kulikov S and Hołderna-Natkaniec K 2005 *Proc. ICANS-XVII*, Santa Fe, ed G J Russell, J J Rhyne and B V Maes Los Alamos Report LA-UR-06-3904 **Vol. II** 519-529.
[6] Gundorin A N and Nazarov V M 1980 *Comm. JINR* 3-80-721, Dubna.
[7] Kazimirov V Yu and Natkaniec I 2003 *Comm. JINR* P14—2003-48, Dubna.
[8] Lerf A, He H, Forster M and Klinowski J 1998 *J. Phys. Chem. B* **102** 4477.
[9] Buchsteiner A, Lerf A, Pieper J 2006 *J. Phys. Chem. B* **110** 22328.
[10] Sheka E F, Natkaniec I, Rozhkova N N and Hołderna-Natkaniec K 2014 *JETPletters* **99** 754-759.
[11] Drużbicki K and Natkaniec I, *Chemical Physics Letters* 2014 **600** 106–111.
[12] Szostak M M, Piela K, Hołderna-Natkaniec K, Natkaniec I and Bidzińska E 2014 *Carbohydrate Research* **395** 29–37.
[13] Ivanov A I, Jimenez-Ruiz M and Kulda J 2014 J. Phys. Conf. Ser. in press.

# Recent and future developments on TOSCA at ISIS

**Stewart F. Parker,[1] Felix Fernandez-Alonso,[1,2] Anibal J. Ramirez-Cuesta,[1,3] John Tomkinson,[1] Svemir Rudic,[1] Roberto S. Pinna,[4] Giuseppe Gorini[4] and Javier Fernández Castañon.[5,6]**

[1] ISIS Facility, STFC Rutherford Appleton Laboratory, Chilton, Didcot OX11 0QX, UK
[2] Department of Physics and Astronomy, University College London, Gower Street, London WC1E 6BT, UK
[3] Present address: CEMD, Neutron Sciences Directorate, Oak Ridge National Laboratory, Oak Ridge TN 37831, USA
[4] CNISM, Università degli Studi di Milano-Bicocca, Piazza della Scienza 3, 20126 Milano, Italy
[5] Facultad de Ciencias, Universidad de Oviedo, Calvo Sotelo, 33007 Oviedo, Spain
[6] Present address: ILL, 6 rue Jules Horowitz, BP 156, F-38042 Grenoble Cedex 9, France

E-mail: stewart.parker@stfc.ac.uk

**Abstract**. TOSCA is a high-resolution neutron spectrometer at the ISIS Pulsed Neutron and Muon Source. The instrument is optimised for broadband vibrational spectroscopy in the 0 – 4000 $cm^{-1}$ region and it has been operational since 2000. This paper describes how the instrument has been progressively upgraded in the intervening years to enable new science. Future upgrades are outlined.

## 1. History

An indirect-geometry inelastic neutron scattering (INS) spectrometer optimised for vibrational spectroscopy in the 0 – 4000 $cm^{-1}$ region has been an integral part of the instrument suite at ISIS since the start of neutron beam operations at the facility in 1985 [1]. The first-generation instrument, TFXA, introduced both time and energy focussing to provide unprecedented resolution across the entire energy range [2]. Over the period April 1985 to February 1998, TFXA underwent several minor upgrades that resulted in lower backgrounds and slightly improved resolution.

By 1998, it had become apparent that there was a need for an instrument with greater sensitivity and improved resolution [3], objectives that were subsequently realised in the TOSCA project. The project was jointly funded by CNR (Italy) and HEFCE (UK). The spectrometer was installed in two stages: TOSCA-I had slightly better resolution than TFXA but a larger detector area and hence count-rate [4]. TOSCA-II (hereafter TOSCA) as installed in 2000, is shown in Figure 1, had both improved resolution, by virtue of the primary flight path having been increased from 12 m to 17 m, Figure 2, and sensitivity, by the installation of additional detector banks in forward scattering [2,5].

In its present incarnation, TOSCA is located at the N8 beamline on ISIS Target Station I. It has been operational since 2000. The spectrometer sits at 17 m from a room temperature water moderator

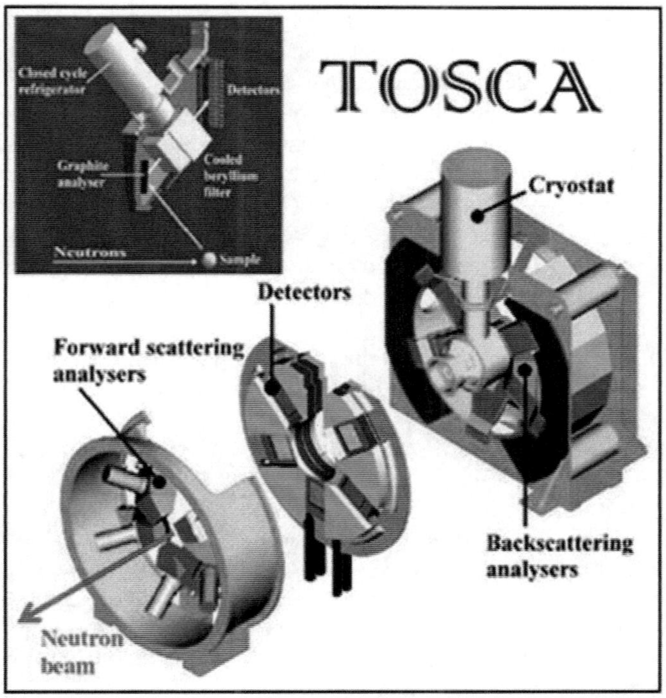


**Figure 1.** Schematic diagram of TOSCA. The inset shows a cross sectional view of a detector module.

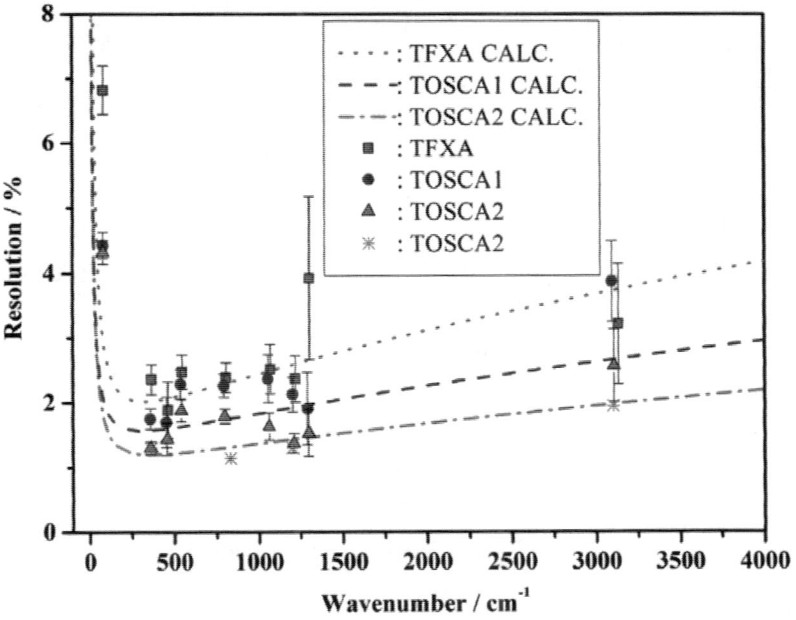


**Figure 2.** Evolution of the spectral resolution on TFXA, TOSCA-I, and TOSCA-II. The lines correspond to analytical estimates.

and it is illuminated by a pulsed polychromatic beam of neutrons. Scattered neutrons are Bragg reflected from a pyrolytic graphite analyser and higher-order reflections beyond (002) are suppressed by a cooled Be filter so as to define a unique final energy of ~32 cm$^{-1}$ (see inset in Figure 1). The detector array is comprised of a total of ten banks each having thirteen 30 × 1 cm$^2$ $^3$He tubes. Five banks are located in forward scattering (scattering angle 45°) and five in backscattering (135°). The use of a low (fixed) final energy translates into a direct relationship between energy transfer ($E_T$, cm$^{-1}$) and momentum transfer ($Q$, Å$^{-1}$) such that $E_T \approx 16Q^2$. A disc chopper to prevent frame overlap is positioned at 8 m from the moderator. TOSCA has been optimised to deliver an outstanding spectral resolution (*ca.* 1.25% $E_T$), arising from the combination of several factors: the narrow bandpass of the PG002/Be analyser, tight moderator pulse widths (~10 μs/Å), a long incident flight path and a time- and energy-focused detector geometry.

The primary aim of this paper is to describe the improvements made to TOSCA since the first neutrons in September 2000, as well as to outline possible developments in the future.

## 2. Upgrades to TOSCA since 2000

### *2.1. Reduction of background*

Following its commissioning in 2000, it was found that TOSCA performed as expected in terms of improved resolution and count rate [5]. However, the background was larger than either expected or desired. For organic compounds, the scattering was strong enough that the resulting signal-to-background ratio was high enough that the spectra were acceptable. However, for non-hydrogenous materials, this was not the case and the spectral features were comparable to the background. This

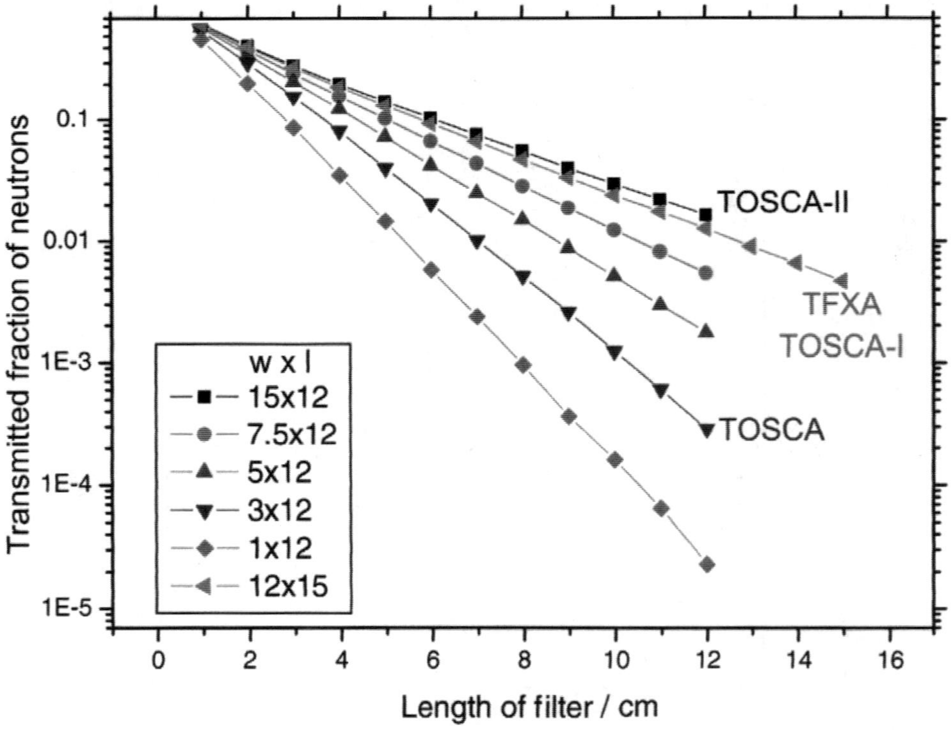


**Figure 3.** Transmission of higher-order neutrons through the beryllium filter as a function of the width (w) and length (l) of the filter, in centimeters. The filter sizes used for TFXA, TOSCA-I and the initial (TOSCA-II) and final implementation (TOSCA) are indicated.

Journal of Physics: Conference Series **554** (2014) 012003 doi:10.1088/1742-6596/554/1/012003

problem was further aggravated by the realisation that the background was sample dependent, thus attempts to run a reference sample and subtract the resulting background were fruitless. The shielding around the sample environment tank, the graphite analysers and the detectors was systematically improved but without any significant reduction in the background.

The problem was finally solved by a series of Monte Carlo simulations of the transmission of the beryllium filter, the results of which are shown in Figure 3. This shows the fraction of higher-order neutrons ($\lambda/2$, $\lambda/3$…, with energies 128, 288… $cm^{-1}$) that pass through the filter as a function of the length and width of the filter. These calculations assume perfectly absorbing sides, a condition that is typically achieved *via* the use of cadmium up to epithermal energies. As such, they represent best-case estimates relative to actual performance in the laboratory, where other secondary sources of background might be possible.

Notwithstanding these limitations, the neutron-transport simulations show that the longer the filter or the narrower the filter, the better the rejection of the higher order neutrons. This explains one of the most puzzling aspects of the problem: why did this not occur with either TFXA or TOSCA-I? There are two factors that are relevant to this discussion: firstly, both used 15 cm long filters, so the rejection was about a factor of four better than the 12 cm long filters used on TOSCA due to space constraints. Secondly, and very fortuitously, the undesired neutrons manifest at about the same time-of-flight as the elastic line when the instrument is at 12 m, thus do not greatly influence the spectrum. With the instrument at 17 m, the 'bad' neutrons still occur at about the same time-of-flight because the incident flight time is not greatly increased for these relatively high-energy neutrons as they must have an energy of at least 128, 288… $cm^{-1}$ *after* scattering, but this time period now corresponds to energy transfers in the 'spectral' range and contaminate the data. This also explains why the background is sample dependent.

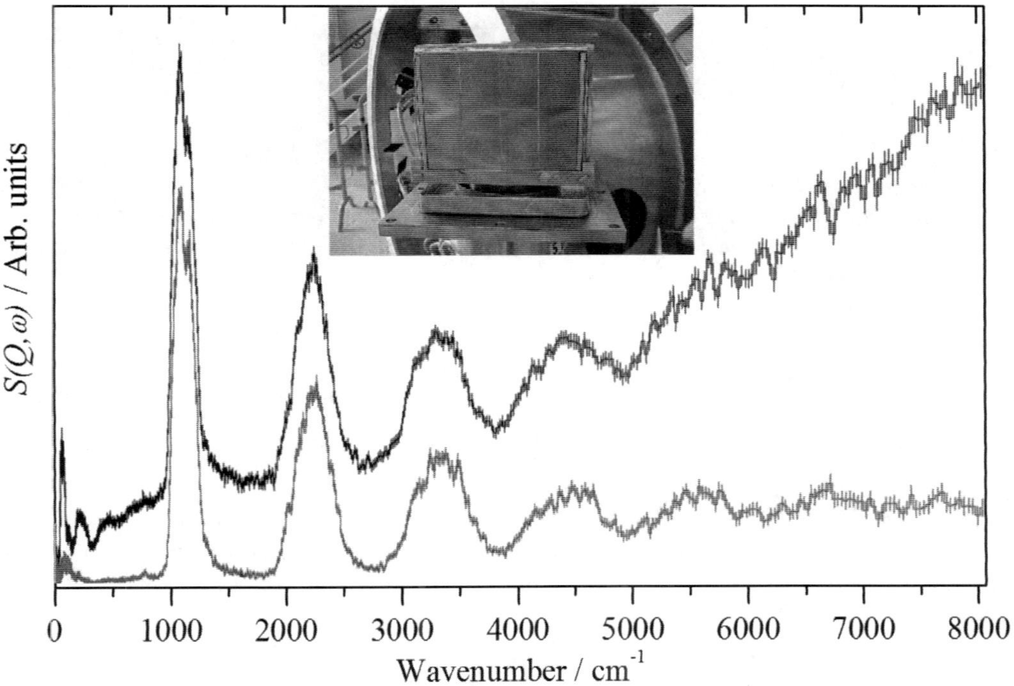


**Figure 4.** Comparison of the INS spectra of $ZrH_2$ before (upper, black) and after (lower, red) the filters were sectioned. The inset shows the beryllium filter after sectioning, the thin vertical lines are the cadmium absorbers.

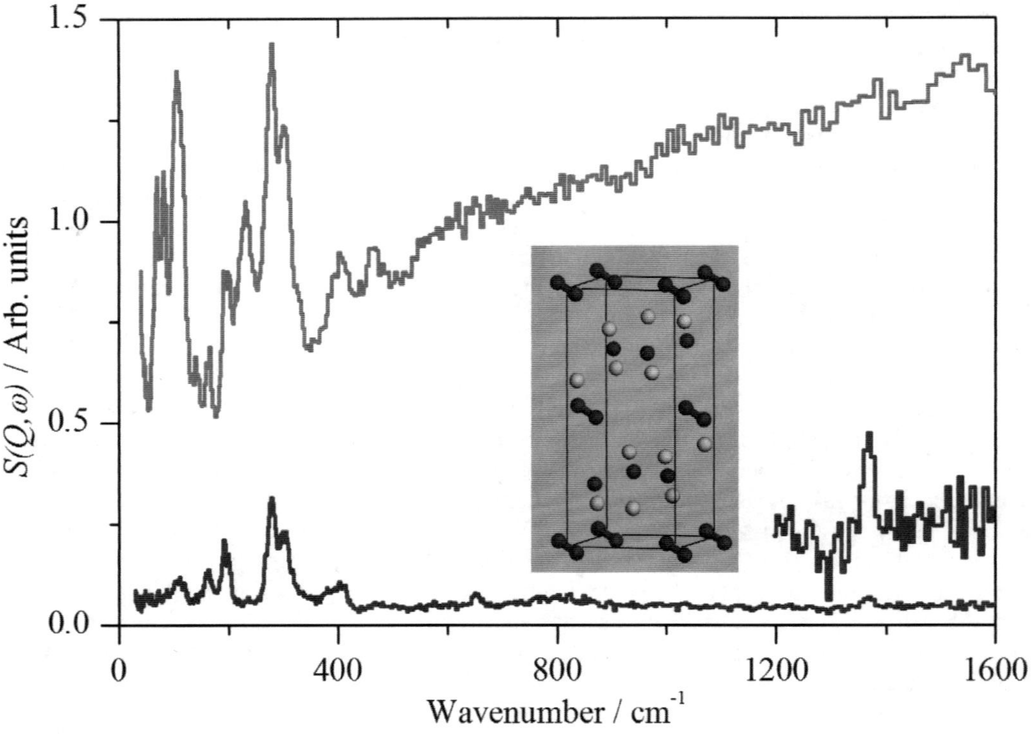


**Figure 5.** Comparison of the INS spectra of SrN [6] recorded before (upper) and after (lower) the filters were sectioned. The region 1200 – 1600 $cm^{-1}$ is shown ×10 ordinate expanded. The crystal structure is shown as an inset (blue = nitrogen, green = strontium).

Previous work [6] has shown that the problem can be effectively addressed by 'poisoning' the filters. The effective width of the filter was reduced by slicing the beryllium into 3 cm wide blocks along the direction of flight and inserting 1 mm sheets of cadmium between the beryllium slices. The results are startling, as shown in Figures 4 and 5. These compare INS spectra of the same sample recorded before and after the filters were sectioned. In Figure 4, both spectra were recorded simultaneously and compares two banks of detectors; one where the beryllium was a solid block (upper, black) and the other where the beryllium had been sectioned (lower, red). The reduction in background by almost a factor of 10, with no loss of intensity in the spectral features is obvious. The noise level is lower for the sectioned case (compare the size of the error bars in the black and red traces) because the unwanted background neutrons contribute to the noise level.

For the non-hydrogenous compound SrN [7] (better described as $\{[Sr^{2+}]_4\ [N^{3-}]_2\ [N_2^{2-}]\}$), the decreased background allows observation of the N=N stretch at 1380 $cm^{-1}$. It also shows that most of the features below 200 $cm^{-1}$ are spurious, greatly simplifying the analysis. The lessons learnt over the course of this upgrade of the TOSCA energy analysers have been instrumental in the specification and subsequent design of a new beryllium filter on the OSIRIS spectrometer at ISIS, a project which was recently successfully completed [8].

### *2.2. Extension of the spectral range*

The installation of TOSCA at 17 m greatly improved the spectral resolution but it restricted the energy transfer range to > 25 $cm^{-1}$. This was because the elastic line occurs at ~23000 μs and since ISIS operates at 50 Hz, a neutron pulse occurs every 20000 μs (a frame), thus fast neutrons would overtake

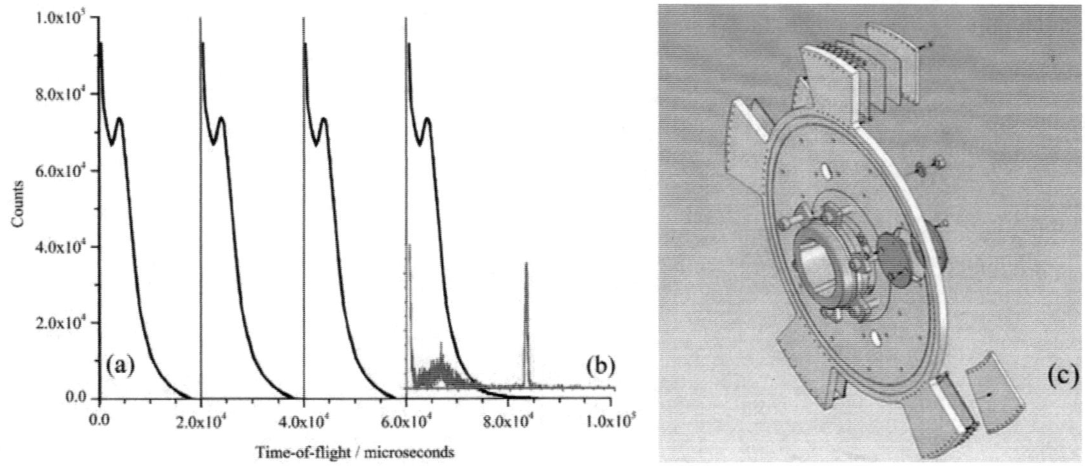


**Figure 6.** (a) Incident monitor spectra (black trace) when TS2 is operating showing the three short 20000 μs frames and the long 40000 μs frame, (b) time-of-flight spectrum (red trace) in an inelastic detector, showing the elastic line at ~23000 μs and (c) schematic of the new disc chopper that enables the long (TS2) frame to be exploited to extend the energy transfer range of TOSCA.

slow neutrons from the preceding pulse. This is catastrophic for time-of-flight analysis since it renders the 'time stamping' that the method depends upon unreliable. In the original form of TOSCA this was prevented by the addition of a 'tail cutter', a sheet of neutron absorbing material on the leading edge of the Nimonic chopper used for suppression of the unmoderated neutrons and γ-rays produced by the proton beam impacting the target. (There was concern that these would cause increased background).

In 2008 ISIS, began operating the second target station (TS2). TS2 takes one pulse in five from the accelerator, resulting in three 20000 μs frames followed by a 40000 μs frame, Figure 6a. Soon thereafter, it was realised that TS2 operation would enable access to energy transfers below 25 $cm^{-1}$, including the elastic line, as shown in Figure 6b. As investigation had shown that the fears concerning background from the prompt pulse were phantasmal, the Nimonic chopper was replaced by a disc chopper, to block the slow neutrons for three pulses and allowed them to pass during the long (TS2) pulse. Note that the spectral range is extended *without* loss of flux. The chopper is shown schematically in Figure 6c.

The result of the innovation is that the energy transfer range now extends to -25 $cm^{-1}$ and includes the elastic line. This development is potentially very useful, as it provides a direct measure of the quantity of orthohydrogen present in a sample when dihydrogen is being studied. The extension to lower energy is illustrated in Figure 7 that compares INS spectra in the low energy region of γ-picoline (4-methylpyridine) recorded on TOSCA and on the high-resolution backscattering spectrometer IRIS. The rotational tunneling transition at 4.2 $cm^{-1}$ is clearly seen in both datasets, the greater resolution of IRIS enables the fine structure of the transition to be observed, (inset at top right of Figure 7).

This development provides overlap between the instruments and potentially allows TOSCA to be used for quasielastic neutron scattering studies of very fast processes, albeit only at two momentum transfer values corresponding to the forward and backscattering detectors. Note that this development is not at the expense of reducing the energy transfer range: the complete spectrum from -25 $cm^{-1}$ to >4000 $cm^{-1}$ is measured.

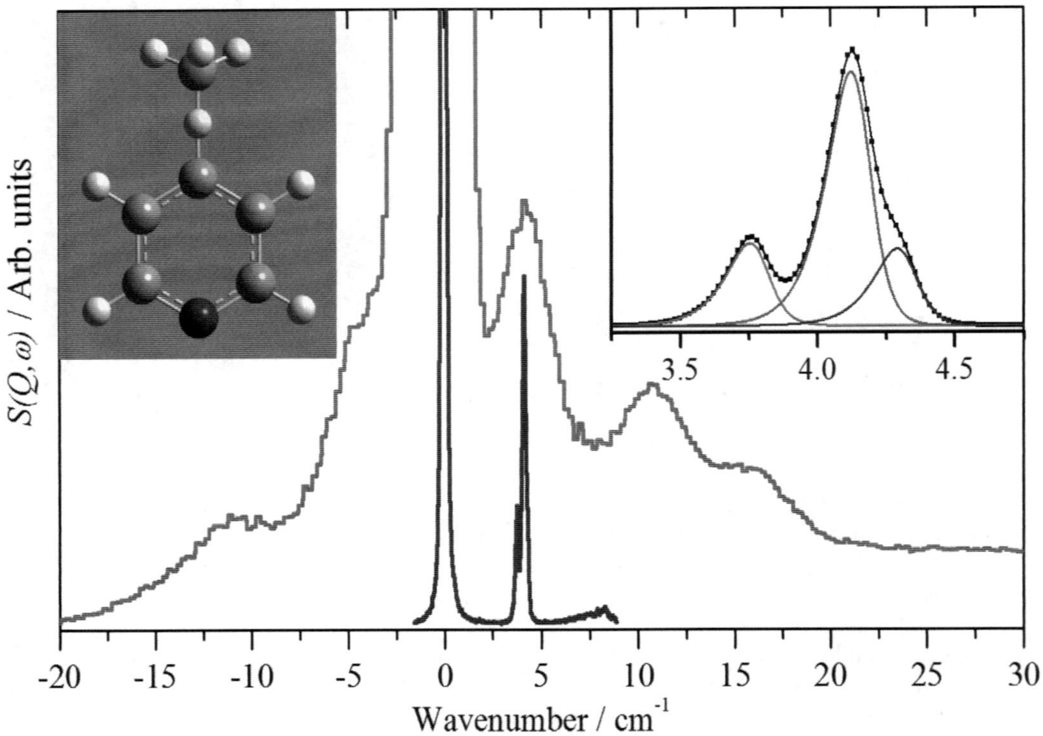


**Figure 7.** Comparison of the INS spectra of γ-picoline (4-methylpyridine, see inset in top left of the figure) recorded on TOSCA (red trace) and on IRIS (blue trace). The rotational tunneling transition at 4.2 cm$^{-1}$ is clearly seen in both datasets, the greater resolution of IRIS at low energy transfers enables the fine structure to be observed, (inset at top right).

### 3. Ongoing and future developments

Neutron guides provide a means to transport neutrons across tens of metres with greatly reduced losses as compared to a simple tube. Mathematically, the process is completely analogous to light transmission along a fibre optic cable except that the transmission occurs for neutrons by total external reflection rather than by total internal reflection as it does for light. When TOSCA was constructed, consideration was given to incorporation of a guide. However, at that time neutron guides typically had $m \approx 1$ (where $m$ is the ratio of the critical angle of the supermirror to that of nickel) and for the relatively hot neutrons required for vibrational spectroscopy, the small gains at low energies did not justify the investment.

In the last decade, both guide technology and the ability to simulate them have advanced significantly. For TOSCA to move forward beyond the current state-of-the-art in key scientific areas such as gas and charge storage, the most urgent need is for greater sensitivity *via* provision of a guide in the primary spectrometer. Simulations of incident flux gains for a series of guide-reflectivity factors are shown in Figure 8. For example, a seamless $m = 5$ guide leads to a *ca.* ×20 increase in flux at the $H_2$ rotational line transition energy (~120 cm$^{-1}$), while a ×3 gain is still attainable at 2000 cm$^{-1}$.

Such an increase in incident flux implies an order-of-magnitude reduction in counting times, a transformational development as it would enable routine studies of industrially relevant systems containing weak neutron scatterers ($CO_2$, $SO_2$, CO, NO) or minute amounts of hydrogen (*e.g.*, proton

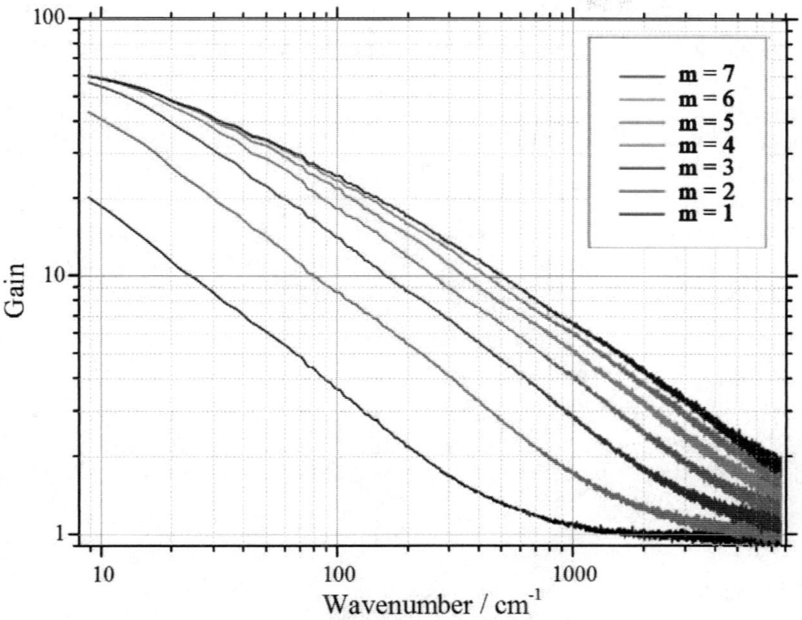


**Figure 8.** Predicted gains in incident flux at the sample position of TOSCA as a function of energy transfer and *m* number.

conductors or supported catalysts). Parametric studies would become routine as well, particularly for hydrogen-containing molecules such as hydrocarbons. This upgrade would also enable access to smaller specimens. At present, 100 mg of a hydrogenous organic sample or several grams of a non-hydrogenous sample remain challenging. A guide should reduce these limits by over an order of magnitude. Likewise, the remit of the (very popular) TOSCA Xpress service could be expanded. These proposals were recently endorsed by an International Review Panel who recommended: '*The panel agreed that the guide upgrade has to be taken up with the highest priority*'.

TOSCA has had modest high-resolution diffraction capabilities since its installation (four $^3$He tubes in backscattering). However, these have not been used extensively because of low count rates and also because coherent scattering is typically swamped by incoherent backgrounds in hydrogenous systems. Given the increasing interest in non-hydrogenous materials, the higher incident flux from a guide would accelerate this on-going trend, also paving the way for much needed simultaneous structural and spectroscopic studies of complex materials under realistic conditions. To this end, additional diffraction modules at 175°, 135° and 45° would significantly (and usefully) extend the *d*-spacing range currently available.

## 4. Conclusions

For the last decade or so, TOSCA has set the standard internationally for broadband, high-resolution vibrational spectroscopy with neutrons. So far, it has out-performed all analogous spectrometers both in terms of spectral resolution and signal-to-background ratio. Across the fingerprint region (800 – 1600 cm$^{-1}$) of the mid-infrared spectral range (400 – 4000 cm$^{-1}$), the resolution of TOSCA has been unmatched by any instrument at ISIS and beyond until relatively recently. Our experience is that in virtually all cases, intrinsic (*e.g.*, heterogeneously broadened) spectral line widths are broader than the instrumental resolution. After a decade of operation, less than half-a-dozen samples have been resolution limited on the instrument.

TOSCA has been such a success that it is the inspiration for VISION [9] at SNS. VISION will have a *ca.* ×100 advantage in detected flux over TOSCA. The similarity of design means that it should also have negligible backgrounds. In addition, the new instrument IN1-Lagrange [10] at ILL delivers a much higher flux than TOSCA (and possibly even more than VISION). However, the resolution remains comparable to TFXA (*i.e.,* similar to that of most currently operating instruments). IN1-Lagrange is probably the spectrometer of choice for very small samples (a few mg or less) and extensive parametric studies, particularly over narrow energy transfer ranges.

The implementation of a guide on TOSCA combined with the planned upgrade in the first target station at ISIS will deliver an instrument that will continue to be at the forefront of vibrational spectroscopy with neutrons in the foreseeable future.

**Acknowledgements**
The authors gratefully acknowledge the UK Science & Technology Facilities Council (STFC) for financial support and access to beamtime at ISIS. This work has been partially supported within the framework of past and current CNR-STFC agreements for collaborative research between Italy and ISIS.

**References**

[1] Penfold J, Tomkinson J, Lobo M and Davidson I, 1985 in: *Proceedings of the Eighth Meeting of the International Collaboration on Advanced Neutron Sources, 8-12 July 1985*: *Volume II, Neutron-scattering instruments*; Editor George C. Stirling. Rutherford Appleton Laboratory report RAL-85-110-2.
[2] Mitchell P C H, Parker S F, Ramirez-Cuesta A J and Tomkinson J 2005 V*ibrational spectroscopy with neutrons, with applications in chemistry, biology, materials science and catalysis* (Singapore: World Scientific)
[3] Parker S F, Carlile C J., Pike T, Tomkinson J, Newport R J, Andreani C, Ricci F P, Sachetti F and Zoppi M 1998, *Physica B* **241-243** 154-156.
[4] Bowden Z A, Celli M, Cilloco F, Colognesi D, Newport R J, Parker S F, Ricci F P, Rossi-Albertini V, Sacchetti F, Tomkinson J and Zoppi M 2000, *Physica B* **276-278** 98-99.
[5] Colognesi D, Celli M, Cilloco F, Newport R J, Parker S F, Rossi-Albertini V, Sacchetti F, Tomkinson J and Zoppi M 2002, *Appl. Phys. A* **74 [Suppl.]** S64–S66.
[6] Webb F J 1969, *Nucl. Instr. Meth.* **69** 325-329,
[7] Auffermann G, Prots Y, Kniep R, Parker S F and Bennington S M 2002, *ChemPhysChem* **3** 815-817.
[8] http://www.isis.stfc.ac.uk/instruments/osiris/be-filter-project/osiris_frd_outline-proposal6678.pdf
[9] http://neutrons.ornl.gov/vision/
[10] https://www.ill.eu/instruments-support/instruments-groups/instruments/in1-taslagrange/description/instrument-layout/

# The use of direct geometry spectrometers in molecular spectroscopy

**Stewart F. Parker,[1] Anibal J. Ramirez-Cuesta,[2] Peter W. Albers[3] and David Lennon[4]**

[1] ISIS Facility, STFC Rutherford Appleton Laboratory, Chilton, Didcot, OX11 0QX, UK
[2] CEMD, Neutron Sciences Directorate, Oak Ridge National Laboratory, Oak Ridge TN 37831 USA
[3] AQura GmbH, D-63457 Hanau, Germany
[4] Department of Chemistry, Joseph Black Building, University of Glasgow, Glasgow, G12 8QQ, UK

E-mail: stewart.parker@stfc.ac.uk

**Abstract**. The advantages and disadvantages of the use of direct geometry spectrometers for molecular spectroscopy and catalysis studies are described. We show that both direct and indirect geometry INS spectrometers are important tools for the study of industrially relevant areas such as catalysis, proton conductors and gas separation. We propose a novel hybrid instrument, Cerberus, that would offer high sensitivity and high-to-reasonable resolution across the entire 'mid-infrared' spectral range that would effectively advance research in these areas.

## 1. Inelastic neutron scattering spectroscopy

Inelastic neutron scattering (INS) spectroscopy [1] is a form of vibrational spectroscopy that is complementary to the conventional methods of infrared and Raman spectroscopies. Its usefulness derives from the properties of the neutron, which is a neutral particle of mass 1.00866 amu, almost identical to that of the hydrogen atom, 1.00794. The neutron has a magnetic moment, which has resulted in it becoming an essential tool for fundamental studies of magnetism [2]. Neutrons are scattered by atomic nuclei *via* the strong force and since the nucleus is only ~0.1% of the diameter of the atom, neutrons are highly penetrating. The nuclear scattering also means that there are no symmetry-based selection rules, so in principle all modes are allowed in INS spectroscopy.
The intensity, $S$, of an INS transition is given by [1]:

$$S(Q,n\omega_i) \propto \frac{(QU_i)^{2n}}{n!}\exp\left(-(QU_{\mathrm{Tot}})^2\right)\sigma \tag{1}$$

Where $\omega_i$ is the $i$th mode at transition energy $\omega$, $n = 1$ for a fundamental, 2 for a first overtone or binary combination, 3 for a second overtone or ternary combination *etc*..., $Q$ is the momentum transfer defined as:

$$\boldsymbol{Q} = \boldsymbol{k}_i - \boldsymbol{k}_f \text{ where } k \equiv 2\pi / \lambda \tag{2}$$

$k$ (Å$^{-1}$) is the wavevector; λ (Å) is the wavelength of the neutron. The dependence of the scattered intensity on, and hence access to, $Q$ is a major distinction between INS spectroscopy and infrared and Raman spectroscopies that only observe modes at $Q \approx 0$. $U_i$ is the root mean square displacement of the atoms in the mode. $\sigma$ is the inelastic scattering cross section of the atom and this is both element and isotope specific and is ~20 times larger for $^1$H than any other nucleus (including $^2$H). Hydrogen is the lightest element, consequently its amplitude of vibration ($U_i$) is the largest of any element. This fact combined with the large cross section means that the scattering from hydrogenous materials is dominated by modes that involve motion of hydrogen. The exponential term in equation (1) is a Debye-Waller factor, $U_{Tot}$ is the total root mean square displacement of all the atoms in all the modes, (both internal and external), and its magnitude is in part determined by thermal motion. This can be reduced by cooling the sample and so spectra are typically recorded below 30 K.

## 2. Indirect and direct INS spectrometers

The first INS spectrometer was the triple axis spectrometer and was invented by Bertram Brockhouse in 1952, as described in his Nobel lecture [3]. This type of instrument, which is still in use today [4], employs monochromators to fix both $\boldsymbol{k_i}$ and $\boldsymbol{k_f}$, thus it is enormously flexible and can (theoretically) access any point in $(Q,\omega)$ space. The strength of the instrument is also its Achilles heel: spectral acquisition is by a point-by-point method and is very slow. The throughput can be greatly increased by fixing either $\boldsymbol{k_i}$ or $\boldsymbol{k_f}$ and using a 'white' beam in the scattered or incident neutrons respectively. Instruments that fix $\boldsymbol{k_i}$ or known as direct geometry instruments and those that fix $\boldsymbol{k_f}$ as indirect geometry instruments.

Historically, indirect geometry instruments have been used for problems that do not involve magnetism. Thus most studies in molecular spectroscopy and catalysis have used such instruments. The reasons for this are instructive. Figure 1 compares the INS, Raman and infrared spectra of N-phenylsuccinimide (the structure is shown as an inset to the figure), a model compound for bismaleimide composites. All three spectra are similar in appearance, so anyone familiar with infrared or Raman spectra can understand the INS spectrum. The complementarity of all three spectra is clear: features that are strong in one spectrum may be weak or absent in the others, thus the out-of-phase carbonyl stretch at 1700 cm$^{-1}$ is very strong in the infrared, weak in the Raman and absent in the INS. Conversely, the out-of-plane carbonyl bend at 407 cm$^{-1}$ is strong in the INS, weak in the Raman and

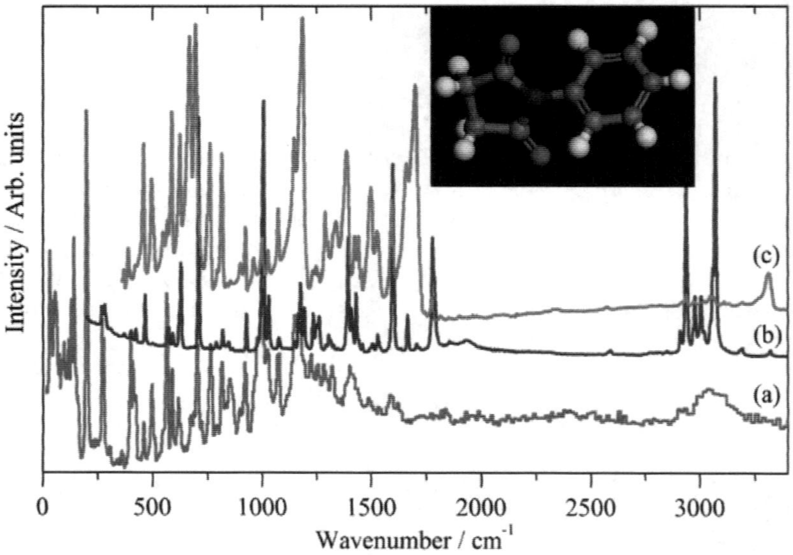


**Figure 1.** Vibrational spectra of N-phenylsuccinimide: (a) INS recorded on TOSCA at 20 K, (b) Raman and (c) infrared at room temperature.

absent in the infrared. The transition energies are the same in all three spectra, since these are a molecular property, however, the intensities differ since the interactions between the probe and the molecule are different in all three cases. The Raman spectrum cuts-off at 200 $cm^{-1}$ and the infrared at 370 $cm^{-1}$ while the INS spectrum extends to almost 0 $cm^{-1}$. This is a region that can be observed by Raman and infrared spectroscopies but it becomes increasingly difficult (and costly) as 0 $cm^{-1}$ is approached. The availability of a wide spectral range is particularly useful in studies of supported metal catalysts where the absorption of light by the support results in a spectral cut-off that can be as high as 1400 $cm^{-1}$. For INS spectroscopy, this is irrelevant and most support materials *e.g.* alumina, silica, carbon, are transparent and the entire spectral range is accessible. Indirect geometry INS spectrometers are generally simple to operate with few user inputs required. For the last decade or so, TOSCA at ISIS has been the world's best spectrometer of this type, this dominance is being challenged by the new instruments VISION at SNS and IN1-Lagrange at ILL. All three instruments are described in detail elsewhere in this volume.

Direct geometry spectrometers have only recently started to be exploited for 'non-magnetism' investigations. There are several reasons why this has occurred. The first is availability. Figure 2 shows the distribution of direct and indirect INS spectrometers and there are significantly more direct geometry spectrometers, particularly those with high resolution and wide spectral range. This pattern is likely to continue, the European Spallation Source has proposed four direct geometry spectrometers and only two indirect geometry spectrometers. (A complete description of the instruments available at neutron facilities can be found *via* http://neutronsources.org/neutron-centres.html).

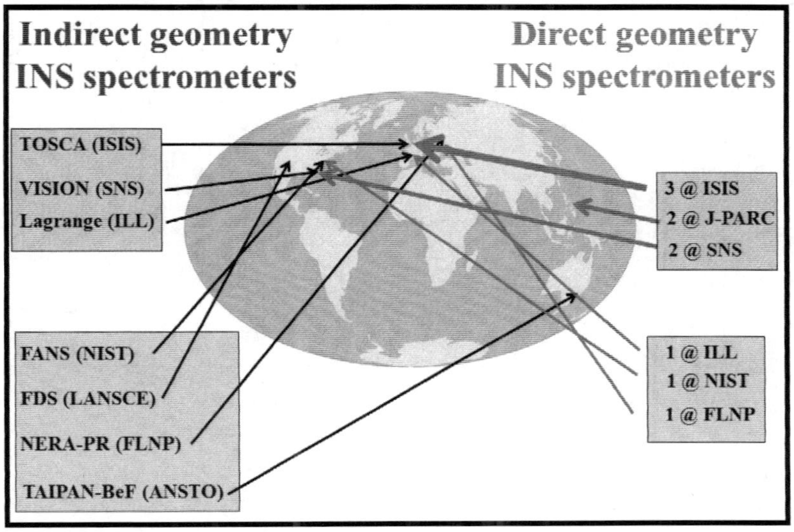


**Figure 2.** Distribution of INS spectrometers. Those in the pink boxes offer high resolution and wide spectral range, those in the blue, lower resolution and/or a restricted spectral range.

A second reason is that direct geometry instruments have large detector areas which makes them very sensitive. Thus MAPS [5] at ISIS has ~16 $m^2$ of detector area, in contrast TOSCA has ~0.4 $m^2$. This results in a significant advantage for MAPS as shown in figure 3. It can be seen that there is a about a factor of three in signal-to-noise ratio in favour of MAPS. If the resolution is relaxed on MAPS [5] then the difference increases to ~10. If lower resolution can be employed then MERLIN [6] at ISIS is almost tenfold more sensitive than MAPS, by virtue of a larger solid angle of detection and greater incident flux by use of a guide, as shown in recent work on hydrocarbon overlayers on

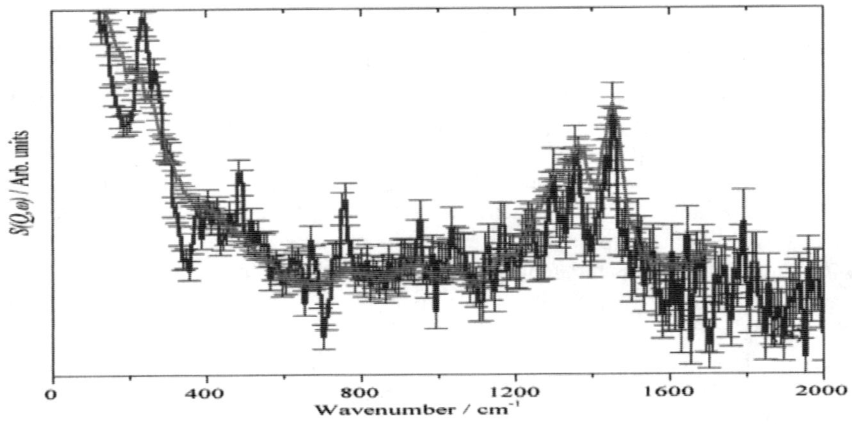


**Figure 3.** Comparison of the spectra recorded on MAPS (red) and TOSCA (blue) from the same sample.

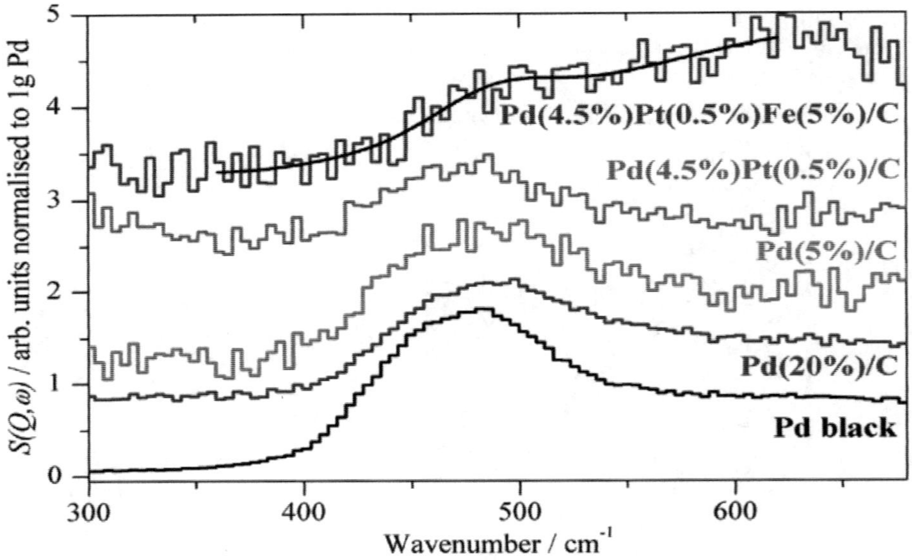


**Figure 4.** Comparison of the INS spectra recorded on MAPS at 20 K of a series of palladium-based catalysts [8] that are relevant to the hydrogenation of nitroarenes to aminoarenes. The peak at 480 $cm^{-1}$ is due to β-PdH. The carbon support and the empty sample can have been subtracted from the spectra.

catalysts [7]. Supported metal catalysts with realistic metal loadings of 1-5 wt% typically have < 1 mol% of hydrogen present, which is the detection limit on TOSCA. The improved sensitivity of the direct geometry instruments means that it is possible to study such systems. An example is shown in figure 4 for a series of palladium-based catalysts [8] that are relevant to the selective catalytic hydrogenation of nitroarenes to aminoarenes for use in polyurethane manufacture. A key parameter is the amount of hydrogen available from the catalyst *via* storage as β-PdH. Even for supported nanoscale ternary alloy particles as found in the industrial PdPtFe/C catalyst, it is still readily detectable at the lowest level.

The sensitivity advantage is likely to become less important with time as the latest generation of indirect geometry spectrometers, VISION at SNS and IN1-Lagrange at ILL, are optimised to have a very large detected flux.

Perhaps the key difference between direct and indirect geometry spectrometers is that the former can access a much larger range in $Q$ and, crucially, can always access smaller $Q$. This is shown in figure 5 which shows the $(Q,\omega)$ plot for iodomethane (methyl iodide, $CH_3I$) recorded with incident energies ($E_i$) of 4840 and 2017 $cm^{-1}$. The purple and green lines show the trajectories of the forward and backscattering banks on TOSCA, however, all low final energy indirect spectrometers follow trajectories that lie either on or between those shown. Thus there is essentially no $Q$-resolution, so it is not possible to scan $Q$ at fixed energy transfer on any operating indirect spectrometer. It can be seen that large energy transfer involves large momentum transfer. The importance of this arises from equation (1), since higher order transitions, *i.e.* $n \geq 2$, depend on $Q^{2n}$ these become significant at large $Q$ and, in particular, phonon wings (combinations between internal modes and the external density of states) become very strong. In parallel there is an overall dampening of intensity caused by the Debye-Waller factor. The combination of these two processes means that the fundamental transitions are largely swamped by the nearby higher order processes, rendering them unobservable. With a direct geometry spectrometer, by only using the detectors in forward scattering *i.e.* at low $Q$, high energy transitions are readily observable. This is illustrated in figure 6, where the data for $4 \leq Q \leq 10$ Å$^{-1}$ obtained from figure 5 is compared to the spectrum of iodomethane recorded on TOSCA. From the TOSCA spectrum, figure 6a, it would be assumed that the arrowed peak at 3043 $cm^{-1}$ were the unresolved C–H stretch modes. Inspection of the MAPS spectrum, figure 6d, has considerably better signal-to-noise ratio and shows a peak at 2948 $cm^{-1}$ (arrowed) that is not apparent in the TOSCA data. In fact, this peak is the unresolved C–H stretch modes, the peaks at 3043, 3161, 3272 $cm^{-1}$ are combinations of the C–H stretch and the intense librational mode at 112 $cm^{-1}$ (CH + lib, CH + 2×lib, CH + 3×lib respectively).

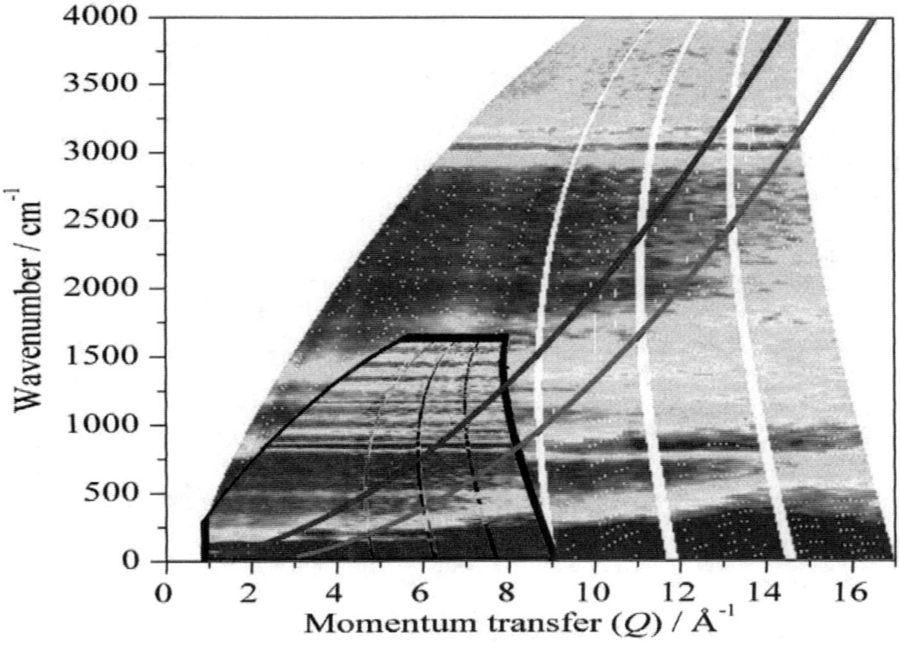


**Figure 5.** $(Q,\omega)$ plot of iodomethane recorded on MAPS at 20 K with incident energies of 4840 and 2017 $cm^{-1}$. The solid lines are the trajectories of an indirect geometry instrument with a final energy of ~40 $cm^{-1}$ for scattering angles of 45° (purple trace) and 135° (olive trace).

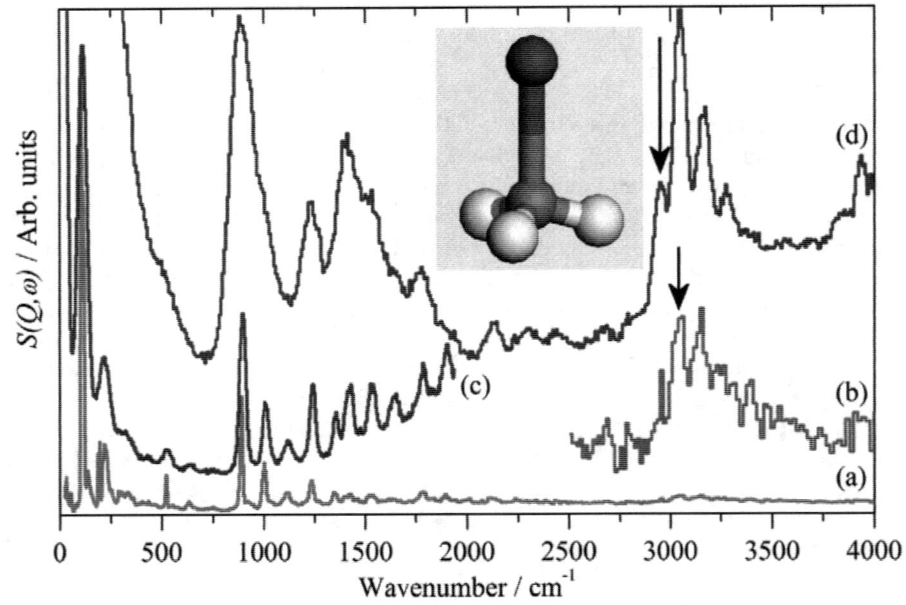


**Figure 6.** INS spectra of iodomethane recorded at 20 K on (a) TOSCA, (b) TOSCA ×10 ordinate expansion of the 2500 – 4000 cm$^{-1}$ region, (c) and (d) MAPS with incident energies of 4840 and 2017 cm$^{-1}$ respectively. (c) and (d) are the data shown in figure 5 summed for the detectors at $4 \leq Q \leq 10$ Å$^{-1}$.

## 3. The next generation - Cerberus

Figures 5 and 6 illustrate the advantages of direct geometry spectrometers: high signal-to-noise ratio, wide $Q$ range and, especially, access to small $Q$. In addition to extending the useful energy transfer range, variable $Q$ also enables higher order processes, $n \geq 2$, to be distinguished [9] from fundamentals $n = 1$, *via* equation (1). Reducing the magnitude of the Debye-Waller factor by using small $Q$ makes experiments at room temperature and above possible [10], where our experience on TOSCA has shown that this is not useful on indirect geometry instruments.

While there are undoubted advantages to direct geometry instruments, there are also drawbacks. The most serious of these is illustrated in figure 6. Practically, for a direct geometry instrument the best resolution is obtained when the energy transfer is close to the incident energy and degrades with decreasing energy transfer, thus for figure 6d, the width of the features around 3000 cm$^{-1}$ is ~60 cm$^{-1}$ (1.2% of $E_i$) whereas for the feature at ~900 cm$^{-1}$ it is 140 cm$^{-1}$ (2.9% of $E_i$). The width of the latter is entirely due to resolution broadening, as the width measured with 2017 cm$^{-1}$ incident energy is 38 cm$^{-1}$ (1.9% of $E_i$) and by TOSCA is 14 cm$^{-1}$. Thus to obtain a high resolution measurement across a wide spectral range, several spectra with different cm$^{-1}$ $E_i$ are needed and these are difficult to merge because of the varying resolution functions. In contrast, the complete spectrum is obtained in one pass with an indirect geometry instrument, in this case the resolution degrades with increasing energy transfer, $E_T$.

There is a further disadvantage with direct geometry instruments, their very sophistication means that they require significant user input. Thus at ISIS the variables are:

**Choice of instrument (×3 @ ISIS)**
**Choice of Fermi chopper slit package (×3 @ MAPS)**
**Choice of Fermi rotation speed (×11 @ MAPS)**
**Choice of incident energy (120 – 16000 cm$^{-1}$ @ MAPS)**

(Similar considerations apply for the SNS and J-PARC). Whereas on TOSCA the only decision is how long to measure for!

It is apparent that there is considerable complementarity between direct and indirect geometry instruments and this is particularly manifested for molecular spectroscopy and catalysis studies. It would be doubly advantageous to combine both types of operation in a single instrument. This would enable studies across the full spectral range to be performed without the need to transfer (or prepare new) samples between instruments, increasing both efficiency and the certainty that the same material is being measured on both instruments.

Figure 7 shows a conceptual diagram for such an instrument – Cerberus. This instrument would be optimised for sensitivity while maintaining reasonable resolution. The instrument would sit at 12 m and view an ambient temperature water moderator with a state-of-the-art neutron guide. For indirect geometry, it would use the large collection area analysers developed on VISION at SNS in 135° backscattering. This would provide a resolution of ~2.25% $E_T$, (*c.f.* TOSCA ~1.25% $E_T$). In forward scattering at 45° would be either more high resolution modules or, to provide sensitivity at the cost of resolution, would be analyser modules that each consist of a cooled beryllium + graphite filter. These would be designed such that they can also operate in a beryllium-only mode to further increase the bandpass and hence flux. The effect on the resolution is shown in figure 8, it can be seen that even at the lowest resolution there is still useful information to be obtained.

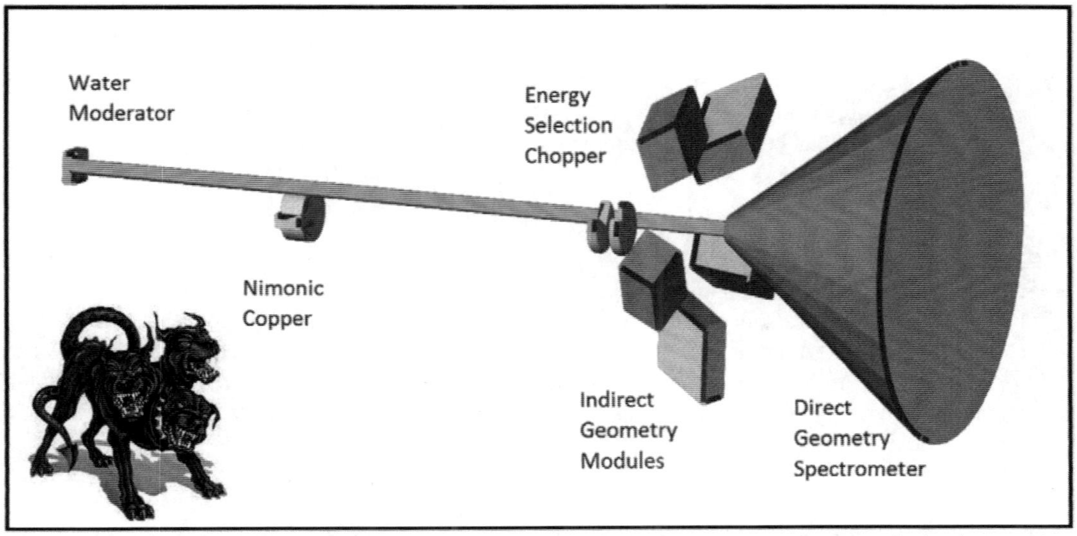


**Figure 7.** The key components of Cerberus showing the indirect geometry detectors in backscattering (blue) and the area detector in forward scattering for use in direct geometry and diffraction.

Also in forward scattering at 2.5 m from the sample will be a large area detector (angular range of ±30° horizontal and vertical) that will be used in conjunction with a Fermi chopper in the incident beam, which will be moved out of the beam when the instrument is in indirect geometry mode, note that measuring spectra in the indirect and direct modes is not possible simultaneously. In indirect mode this detector will function as a low resolution, high count-rate diffraction detector, which will complement a high resolution diffraction detector sited in near-perfect backscattering. The direct geometry forward scattering detectors should offer the same energy resolution as the equivalent area on MERLIN, which we have shown is sufficient to distinguish aliphatic and aromatic C–H stretch modes [7]. To reduce the cost, the detector need not be position sensitive. Experience has shown that $Q$-resolution in the C–H/N–H/O–H stretch region (which is what this detector is needed for) is not essential as all the detectors are summed to improve the signal-to-noise ratio.

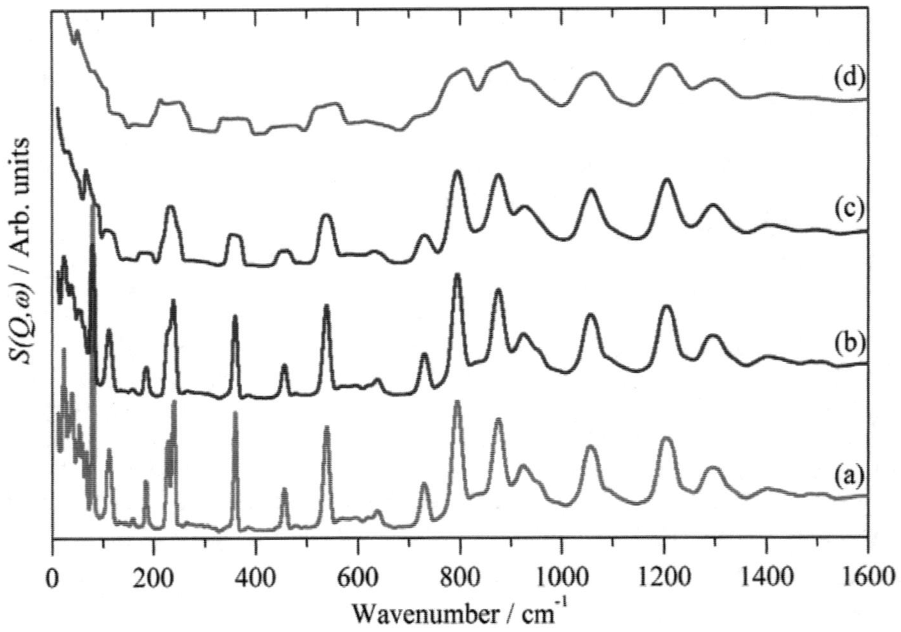


**Figure 8.** Simulation of the effect on the resolution on the INS spectrum of 2,5-diiodothiophene by the use of high throughput detector banks on Cerberus. (a) Experimental TOSCA spectrum, (b) simulated for 12 m incident flight path (high resolution banks), (c) simulated for Be + C banks and (d) simulated for Be only banks.

There are three fields where Cerberus would have an immediate impact: (1) catalysis, (2) proton conductors and (3) gas separation and storage.

(1) Much of catalysis is concerned with the transfer of hydrogen between reagents and products and Cerberus is ideally suited to these type of studies. For the first time it will enable good resolution data to be collected on the same sample across the entire energy range. Access to the C–H/N–H/O–H stretch region provides a very simple means of identification and is ideally suited to quantitation of the species present [11]. The 0 – 2000 $cm^{-1}$ region is where the details of the catalyst-adsorbate interaction can be discerned and the indirect backscattering detectors provide the good energy resolution needed for this. However, the surface species of interest are often present in low concentration and good resolution is of no use if there is insufficient sensitivity. This problem is addressed by the indirect forward scattering which will provide at least a tenfold increase in sensitivity and possibly as much as a 100-fold increase over the backscattering detectors. The improved sensitivity will also allow industrially important non-hydrogenous adsorbates , such as $CO_x$, $NO_x$ and $SO_x$ to be studied by INS for the first time.

(2) Proton conductors are heavy metal oxides that are hydrated to generate mobile hydroxyls and have considerable potential in intermediate temperature fuel cells. These materials exhibit multiple sites for incorporation of hydroxyls, thus diffraction studies are only able to provide an average picture of the material, the crucially important local structure around the hydroxyl groups is lost. Spectroscopy provides this missing local picture and INS is particularly matched to these needs as the low energy modes that (in conjunction with first principles modelling) enable discrimination between the various possible sites are easily observed [12]. These modes are often weak or obscured to optical probes. Access to the O–H stretch region is needed to quantify the species present and to discriminate between hydroxyl groups and water.

(3) Gas separation is becoming an increasingly important activity with the major driver being the need to separate $CO_2$ from waste streams. Recent work on TOSCA has shown that it is possible to detect changes in materials caused by $SO_2$ or $CO_2$ gas absorption [13] but that it is at the limits of sensitivity of the instrument. The increased flux of Cerberus will enable such studies to become routine.

## 4. Conclusions

In this paper we have highlighted the advantages and disadvantages of the use of direct geometry spectrometers for molecular spectroscopy and catalysis studies. To date, these instruments have not been extensively used for such studies but they are likely to comprise an increasingly important component of the workload on such instruments. At ISIS this is already happening; currently on MAPS ~25% of beamtime is non-magnetism, whereas five years ago it was almost zero.

It has become clear that both direct and indirect geometry INS spectrometers are important tools for the study of industrially relevant areas such as catalysis, proton conductors and gas separation. We strongly believe that these areas could be very effectively addressed by a novel hybrid instrument, Cerberus, that would offer high sensitivity and high to reasonable resolution across the entire 'mid-infrared' spectral range.

## Acknowledgments

The STFC Rutherford Appleton Laboratory is thanked for access to neutron beam facilities. The EPSRC are thanked for project support (Grant EP/E028861/1).

## References

[1] Mitchell P C H, Parker S F, Ramirez-Cuesta A J and Tomkinson J 2005 V*ibrational spectroscopy with neutrons, with applications in chemistry, biology, materials science and catalysis* (Singapore: World Scientific).

[2] Perring T G 2009 *Materials Today* **12**, 100-103.

[3] Brockhouse B. N. 1994 Nobel Lecture: Early development of neutron scattering, http://www.nobelprize.org/nobel_prizes/physics/laureates/1994/brockhouse-lecture.html

[4] Shirane G, Shapiro S M and Tranquada J M 2002 *Neutron Scattering with a Triple-Axis Spectrometer*, (Cambridge: Cambridge University Press).

[5] Parker S F, Lennon D and Albers P W 2011 *Applied. Spectroscopy* **65** 1325-1341.

[6] Bewley R I, Eccleston R S, McEwen K A, Hayden S M, Dove M T, Bennington S M, Treadgold J R and Coleman R L S 2006 *Physica B – Condensed. Matter* **385–386** 1029.

[7] Lennon D, Warringham R, Guidi T and Parker S F 2013 *Chemical Physics* **427** 49-53.

[8] Möbus K, Grünewald E, Wieland S, Parker S F and Albers P W 2014 *Journal of Catalysis* **311** 153-160.

[9] Parker S F, Bennington S M, Ramirez-Cuesta A J, Auffermann G, Bronger W, Herman H, Williams K P J and Smith T 2003 *Journal of the American Chemical Society* **125** 11656-11661.

[10] Parker S F 2011 *Chemical Communications* **47** 1998-1990.

[11] Silverwood I P, Hamilton N G, Laycock C J, Staniforth J Z, Ormerod R M, Frost C D, Parker S F and Lennon D 2010 *Physical Chemistry Chemical Physics* **12** 3102-3107.

[12] Karlsson M, Matic A, Parker S F, Ahmed I, Börjesson L andEriksson S 2008 *Physical Review B,* **77** 104302.

[13] Yang S, Sun J, Ramirez-Cuesta A J, Callear S K, David W I F, Anderson D P, Newby R, Blake A J, Parker J E, Tang C C and Schroeder M 2012 *Nature Chemistry* **4** 887-894.

# Sample environment issues relevant to the acquisition of inelastic neutron scattering measurements of heterogeneous catalyst samples

**R Warringham[1], D Bellaire[1], S F Parker[2], J Taylor[2], R A Ewings[2], C M Goodway[2], M Kibble[2], S R. Wakefield[2], M Jura[2], M P Dudman[2], R P Tooze[3], P B Webb[3] and D Lennon[1]***

[1] School of Chemistry, Joseph Black Building, University of Glasgow, Glasgow, G12 8QQ, UK
[2] ISIS Facility, STFC Rutherford Appleton Laboratory, Chilton, Didcot, Oxon OX11 0QX, UK
[3] Sasol Technology UK Ltd, Purdie Building, North Haugh, University of St. Andrews, St Andrews, Fife, KY16 9ST, UK

Email: David.Lennon@glasgow.ac.uk

**Abstract**. Refinements to established sample handling issues connected with the acquisition of inelastic neutron scattering (INS) spectra of heterogeneous catalyst samples are described. This involves modification of a reactor and improvements to gas handling apparatus. The latter includes provision of laboratory space and ventilation facilities that are suited to the increasingly wide range of catalytic samples being examined. INS measurements for CO hydrogenation over an iron-based Fischer-Tropsch catalyst recorded using the MARI spectrometer are presented to illustrate the applicability of the new arrangements.

## 1. Introduction

The technique of inelastic neutron scattering (INS) is a useful method for obtaining the vibrational spectrum of 'black' samples that might contain significant quantities carbon and/ or metal, as these materials present significant challenges when conventional optical techniques, such as infrared (IR) spectroscopy, are used [1]. Further, materials such as metal oxides may also exhibit restricted spectral range in IR measurements due to strong absorption by the lattice modes and their overtones and combinations. For example, IR measurements of alumina are typically restricted to the range 1100 – 4000 $cm^{-1}$ [2], whereas this range may be extended to 50-4000 $cm^{-1}$ when INS is used [3]. Against this background, the recent realisation that direct geometry neutron spectrometers, such as MARI and MAPS, provide some advantages in terms of sensitivity, spectral range, and resolution with respect to indirect geometry spectrometers, such as TOSCA, is enabling INS to be applied to a wider range of technically relevant samples than was previously the case [4]. One such class of sample is heterogeneous catalysts.

For many years IR spectroscopy has been used to examine heterogeneous catalysts and chemisorbed systems [5], as it provides a combination of high sensitivity and excellent spectral resolution. Moreover, modern commercial IR spectrometers may be coupled to conventional catalyst test apparatus, facilitating a platform for *operando* spectroscopy [6]. INS spectroscopy supplements that valuable dataset by being able to examine technical grade catalysts post-reaction, something that, for the reasons mentioned above, is not so easily achievable with optical techniques. However, INS is relatively insensitive; therefore, for studies concentrating on the surface chemistry of catalysts or on the form of chemisorbed species, sample masses typically ≥10 g are required, as opposed to the 20-50 mg typically used in IR measurements [7]. This restriction forces INS measurements away from the micro-reactor configuration commonly encountered in laboratory catalyst testing and requires a significant degree of scale-up for INS sample handling apparatus. Presently, the majority of INS studies on catalysts have tended to feature chemistry at the gas/solid interface [1].

A recent collaboration between the universities of Glasgow and Keele and the ISIS Facility of the STFC Rutherford Appleton Laboratory has used INS to investigate a series of alumina-supported nickel catalysts active for the reforming of methane in to synthesis gas ($CO + H_2$). Initial studies using $CO_2$ as the oxidant, *i.e.* 'dry' reforming, utilised an indirect geometry neutron spectrometer (TOSCA) and produced weak INS spectra; demonstrating that these catalysts are very efficient at cycling hydrogen [8]. Subsequent studies were performed on a direct geometry chopper instrument (MAPS), where improved sensitivity yielded better-defined spectra [8-10]. The investigations were extended to examine the more industrially relevant steam reforming reaction of methane; with that work providing further challenges in relation to sample handling capability [11]. These studies forced the development and commissioning of a new catalyst handling apparatus that was constructed at ISIS and is described elsewhere [12]. The adoption of such sample handling procedures is an essential part of being able to successfully apply INS to probe the surface chemistry of high surface area catalytic systems.

Whilst the INS studies of the nickel based catalysts provided new insight into how hydrogen partitions within the catalyst matrix for the economically relevant methane reforming reactions, it also exposed vulnerabilities in the apparatus used to prepare the samples. Namely, the equipment was quite bulky and, consequently, in a busy Central Facility with high demand for laboratory space, it had to be assembled for each experimental run and then disassembled and stored afterwards. The equipment worked well but, when assembled, it took up two fume cupboards in addition to a section of laboratory floor space required to accommodate the various gas cylinders connected with each experiment. With an increasing number of applications for neutron beam time to study catalytic systems at ISIS, this arrangement became untenable. To overcome these issues, a more compact and more permanent arrangement was proposed. This communication describes the resulting improvements in availability of hardware and outlines the associated methodology. Firstly, an improved gas handling system and associated laboratory infrastructure is outlined. Secondly, a new stainless steel reactor is described that is designed to undertake reactions under more modest conditions of temperature and pressure than required in the aforementioned experiments on nickel based methane reforming catalysts. Further, in order to demonstrate the usefulness of this upgrade in capability, the acquisition of INS spectra for CO hydrogenation over an iron-based Fisher-Tropsch (F-T) catalyst is presented.

## 2. Experimental

### 2.1 Gas handling facility

Figure 1 shows a schematic drawing of the new gas handling apparatus. In concept, the layout is similar to that successfully adopted in the initial layout [12]; however, the design has been modified to significantly improve the compactness of the whole apparatus, so that the gas handling section can all be accommodated on a single panel. The panel is portable but resides within a walk-in fume cupboard (Kotterman, type 2-451-JAND, average duct velocity = 0.54 $ms^{-1}$) within a sample preparation area off the main experimental hall of the ISIS Facility. A gas cylinder compound has been constructed outside the laboratory and four gas lines (presently configured for helium, carbon monoxide, hydrogen and carbon dioxide) are piped in to the fume cupboard *via* individual regulators and isolation valves. Components are connected using 1/4 in. o.d. stainless steel tubing and Swagelok tube fittings. The laboratory is fitted with a gas detection system (Gas Measurements Instruments Ltd., Active-8 alarm system) in order to warn the operator of any release of hazardous gases. Specifically, two CO and two hydrogen gas detectors are located alongside the gas line regulators and adjacent to the fume cupboard. A manually activated electrical switch mounted on the side of the fume cupboard can be used to trigger in-line isolation valves mounted above the gas cylinder compound to immediately

 IOP Publishing
doi:10.1088/1742-6596/554/1/012005

isolate the gas supply if required. This combination of features allows higher gas flow rates to be permissible than was previously the case with the temporary arrangement when the gas cylinders were contained within the laboratory. Higher permissible gas flow rates mean that reactions can be studied at gas hourly space velocities and residence times that now more closely resemble values accessible in micro-reactor studies. Gas flow to the reactor is controlled by four Hastings model HFC302 mass flow controllers (MFC) attached to a Teledyne THPS-400 controller. These controllers have been calibrated for nitrogen flow but the use of gas correction factors allows other gases to be accurately metered. One controller has a maximum flow rate of 2000 standard cubic centimeters per minute (sccm) whilst the other three are specified at 200 sccm (all $N_2$).

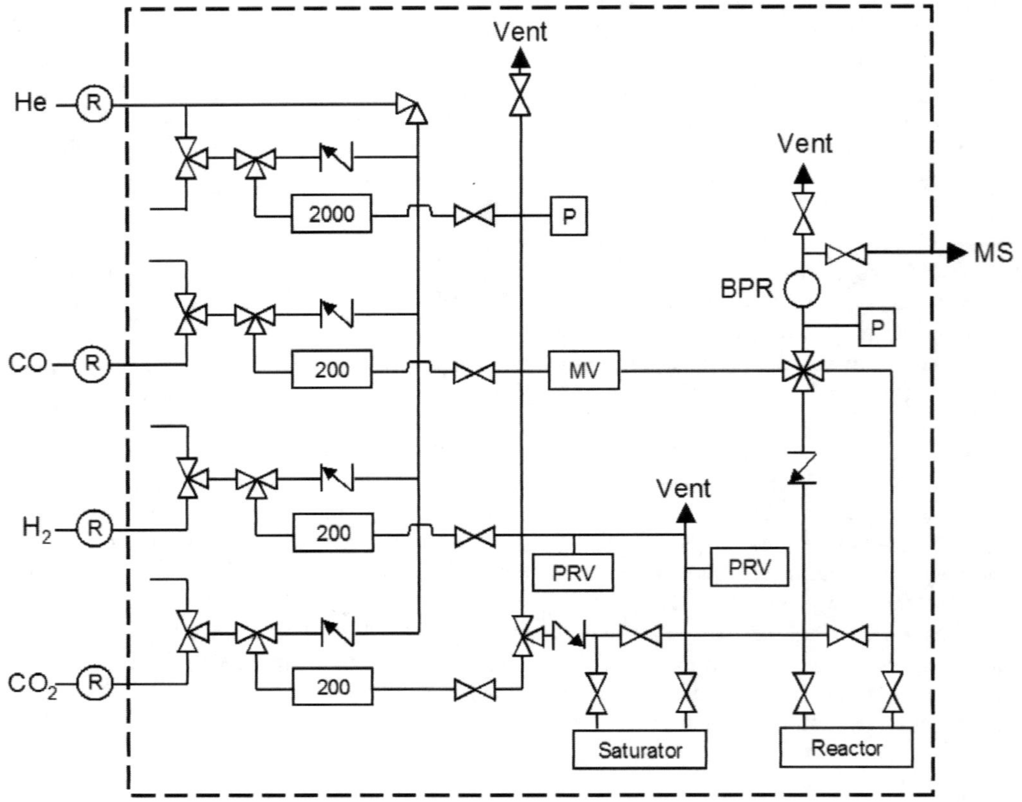


**Figure 1.** A schematic representation of the gas handling apparatus. The mass flow controllers are indicated with the rectangular boxes with the maximum flow rate (sccm) contained within. Other abbreviations include; P – pressure sensor, R – regulator, PRV – pressure relief valve, MV – mixing volume, BPR – back pressure regulator and MS – mass spectrometer. The dashed box schematically represents the enclosed fume cupboard.

A saturator is accessible *via* the inlet line, which means that reactions involving vapourised reagents, *e.g.* steam [11] or methanol [2], can be passed over the catalyst sample. Heated lines connect the gas manifold to the reactor, which is housed within a bucket furnace (Instron SFL, model no: TF105/3/12/F) that is controlled by a PID module (Eurotherm 3508). The furnace is mounted on a heavy duty lab-jack, so that the furnace can be raised to enclose the reactor during reaction and lowered to facilitate disconnection when the cell requires demounting from the gas line. The pressure of the gas line is controlled by a back pressure regulator (Parker Veriflo ABP-1ST-43-PP-X4), with check relief valves placed at several points within the gas manifold. The apparatus is variably rated (see later) up to a maximum operating pressure of 20 bar. Pressure gauges (Druck DPI 280 Series) are located upstream and downstream of the reactor. Pressure readings from these gauges provide information on the presence of any pressure drop in the reactor zone. The reactor exit line can be passed though a 'catch-pot', which will extract condensable products if required. The reactor exit gases are then analysed by mass spectrometry. Initially, this involved a removable quadrupole mass spectrometer (MKS Spectra Microvision Plus) that sampled the eluting gas stream via a fine control valve. That arrangement has now been replaced with a permanently dedicated mass spectrometer (Hiden Analytical, HPR20 QMS Sampling System) that samples the gas stream via a differentially pumped continuous leak. The new arrangement simplifies calibration of the product and reaction streams. The exit lines are all vented within the fume cupboard.

Figure 2 presents a photograph of the gas panel mounted within the fume cupboard. The MFC control unit is located at the front of the blue gas panel. A two-section glass sash at the front of the fume cupboard means that the mass flow controllers can be adjusted by the operator without compromising the negative pressure within the fume cupboard. The furnace is visible in Figure 2 to the left hand side of the gas panel, with the reactor visible above the furnace. The heated lines to and from the reactor are terminated by high temperature valves (Parker HNV54A3) that are capable of operating at temperatures up to 811 K.

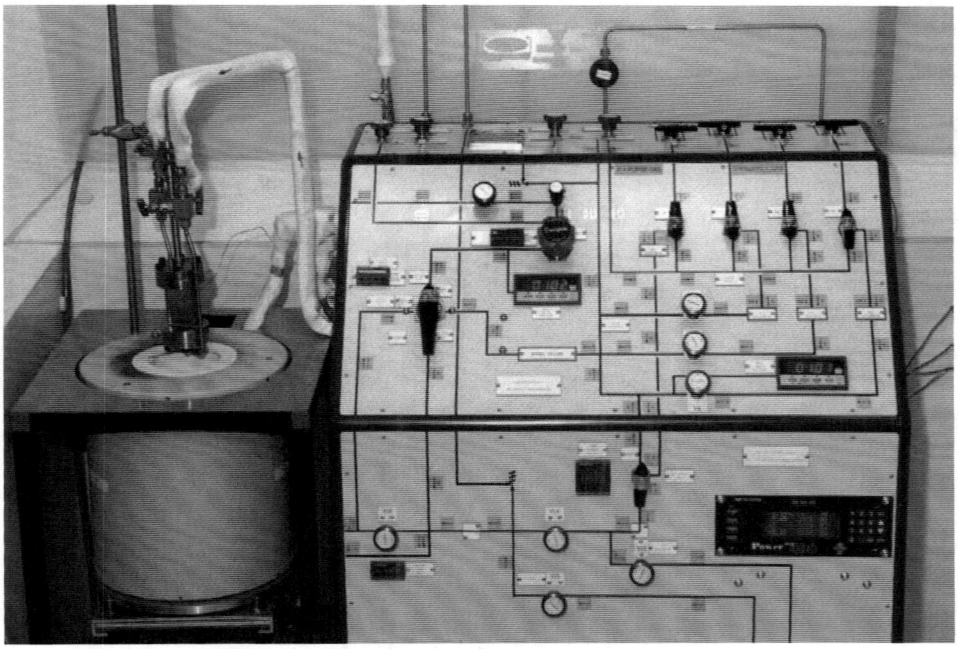

**Figure 2.** A photograph of the gas handling apparatus (blue panel), that is housed within a walk-in fume cupboard. The furnace is located to the left hand side of the gas panel and is mounted on a heavy-duty lab-jack. Above the furnace the reactor can be seen, which is connected to the gas panel by heating lines that terminate in high temperature valves, which may be used to isolate the cell. During operation, the furnace is raised so as to enclose the reaction cell.

## 2.2 Reactor

The Inconel reactor described previously is capable of operation at temperatures up to 873 K and pressures up to 20 bar [12]. It has been used adopting a 'quench' format [13-14] where, post-reaction, the reactor is isolated, transferred to the neutron spectrometer and the spectrum acquired. The specification of the reactor enables high temperature reactions such as methane reforming to be studied in this way [10-11]. However, the Inconel body of the reactor has a vibrational signature that occludes a significant part of the low frequency region of the spectrum [12]; an attribute that constrains the use of the Inconel reactor for spectral acquisition. Furthermore, its significant mass means that it takes long periods of time to cool from reaction temperature to ambient temperature for disconnection from the gas line, and then subsequent cooling to the cryogenic temperatures, *ca.* 20-30 K, typically used for the measurement of the INS spectrum. To address these matters, a new reactor has been constructed from stainless steel. Figure 3 presents a photograph of this reactor, which is intended for lower temperature reactions ($\leq$ 673 K) than obtainable with the Inconel reactor. As with the earlier reactor design [12], the flanges at the bottom and top of the cell are based on Conflat fittings sealed by copper gaskets but now the flanges have been reduced from a thickness of 12 mm to 6 mm, which reduces mass and enables the reactor to be cooled quicker. The lighter reactor makes gravimetric measurements easier and, if desired, it is simpler to mount on a conventional INS sample 'stick' that positions the sample in the cryostat during the INS measurements [15]. The reduction of flange width restricts the operational pressure range of the cell to a maximum operating pressure of 5 bar. Thus, the new reactor is intended for modest temperature and pressure reactions, such as hydrogenations, rather than the more demanding conditions such as encountered with methane reforming. Different check relief valves are fitted to the gas lines depending on the type of reactor being used (Inconel reactor: 20 bar; stainless steel reactor: 5 bar).

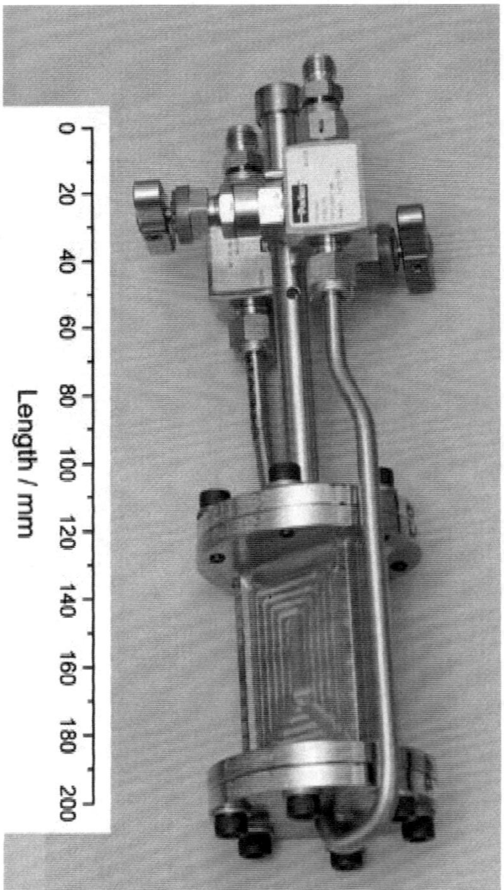


**Figure 3.** Photograph of the stainless steel reaction cell for INS measurements.

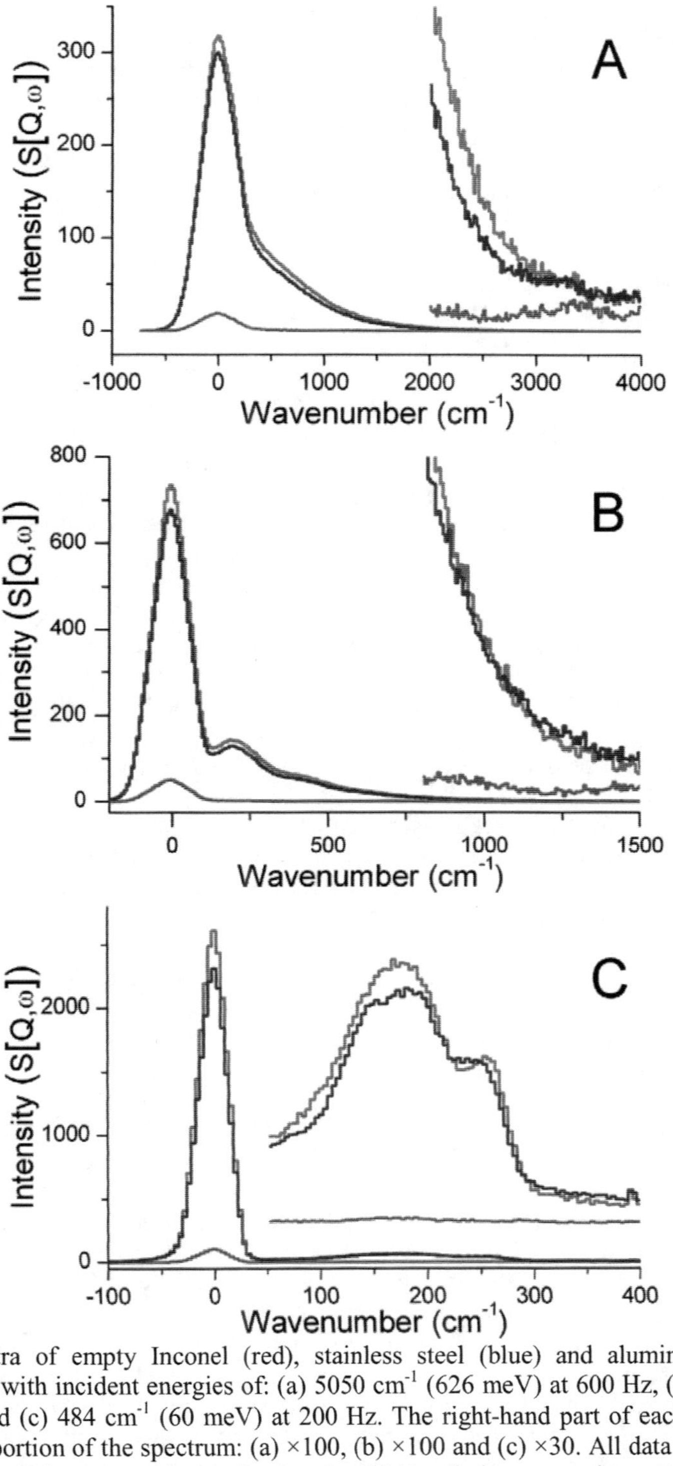


Figure 4. INS spectra of empty Inconel (red), stainless steel (blue) and aluminium (olive) cells recorded on MAPS with incident energies of: (a) 5050 cm$^{-1}$ (626 meV) at 600 Hz, (b) 2016 cm$^{-1}$ (250 meV) at 400 Hz and (c) 484 cm$^{-1}$ (60 meV) at 200 Hz. The right-hand part of each figure shows an ordinate expanded portion of the spectrum: (a) ×100, (b) ×100 and (c) ×30. All data were recorded for 150 μA hr (approx. one hour measurement) using the 'S' Fermi chopper package at the rotation speed given. The different rotation rates largely account for the difference in elastic line intensity between (a), (b) and (c).

In order to compare the relative contributions to the INS spectrum of the two types of reactor, the spectra of an empty Inconel reactor and a stainless steel reactor were recorded at three different incident energies (5050, 2016 and 484 $cm^{-1}$) on the MAPS spectrometer using the S chopper package; the results are presented in Figure 3. Although the stainless steel reactor does present a reduced signal over regions of the spectrum, this difference is minimal. Previous work on the Inconel reactor using the MERLIN spectrometer did demonstrate that, despite a significant background contribution, the reactor was amenable to background subtraction strategies that enabled the INS spectrum of small populations of chemisorbed entities present at a catalyst surface to be obtained [12]. However, from inspection of Figure 3, it is evident that neither reactor is particularly well suited to spectral acquisition purposes. This was not the anticipated outcome envisaged when the stainless steel reactor was commissioned.

To maximize the scattering intensity of surface species present at catalyst surfaces an alternative approach was adopted. Namely, following a procedure used in earlier catalyst studies that utilized a quartz tubular reactor [8-9, 16], the catalyst sample post-reaction was transferred to a conventional aluminum INS sample cell [9] and the INS spectrum recorded. The INS spectrum of an empty aluminum sample holder is also presented in Figure 4 (olive line), where a dramatic contrast to that observed for the denser metal cells is apparent: the aluminum cell exhibits a flat background from 4000 $cm^{-1}$ down to the tail of the elastic peak. So, although aluminum is an unsuitable material for construction of a catalytic reactor due to its relatively low melting point and possible reactivity at elevated temperatures, it is an excellent choice of material for use as an INS sample holder. This scenario reinforces the idea of a separate reactor and INS measurement cell that necessitates the provision of a high specification glove box to facilitate the transfer of sample between the two containers. Future studies will examine the possibility of avoiding the sample transfer from reactor to a separate INS cell. Nevertheless, at present, for the reaction systems examined here, INS inspection of heterogeneously catalysed reactions constitutes a form of *ex situ* analysis.

The fixed bed arrangement of these reactors can potentially lead to problems with inhomogeneous reaction profiles existing across the length of the reactor bed. For instance, it is possible that a discrete reaction front could slowly progress along the length of the bed (65 mm, Figure 3), so that catalyst at the top of the reactor may experience different chemistry to catalytic material located at the bottom of the bed. One way around this issue is to disperse the catalyst with diluent, such as silicone carbide or alumina, which can also reduce the presence of 'hot spots' in the reactor. However, due to the insensitivity of the INS technique, invariably all of the catalyst charge is used in the INS measurements and, hence, the INS spectra represent an average of all of the material within the reactor. If greater sensitivity could be attained, then it would be possible to use INS to analyse different fractions of the bed and, hence, to build up a composite description of the catalyst throughout the whole length of the bed. Within the constraints of the specifications of the present suite of INS spectrometers at ISIS [4], that desirable option is not available at present.(We note in passing that upgrades to MAPS and TOSCA currently in progress, coupled with the proposed ISIS TS1 upgrade, may make this option feasible in the near future.) Reactions that inherently exhibit low conversions are less prone to these problems. For example, XRD and Raman analysis of an iron Fischer-Tropsch catalyst that had experienced 6 h reaction with syngas at 623 K and atmospheric pressure, with samples taken from different regions of a tubular reactor, showed identical diffraction patterns and Raman spectra; indicating a uniformity of catalyst composition and form throughout the whole length of the reactor [16]. In that case, CO conversion was low (*ca.* 1%). In cases where high feedstock conversions are evident, operators are encouraged to be aware of possible problems with sample inhomogeneity. One way round these difficulties is to run the reactions at reduced temperatures but for longer times. However, the relevance of the operating conditions that favour lower conversions need evaluating in separate micro-reactor experiments. Presently, the analysis of heterogeneous catalyst samples remains a non-trivial process.

## 3. Results and discussion

### 3.1 Reaction testing: CO hydrogenation over an iron-based Fischer-Tropsch catalyst

To illustrate the effectiveness of the new apparatus, a preliminary analysis of an iron-based Fischer-Tropsch catalyst that is active for the hydrogenation of CO is presented in this communication. Iron based Fischer-Tropsch catalysts typically originate as hematite ($\alpha$-$Fe_2O_3$) but under steady state reaction conditions the nature and form of the catalyst is changed so that the catalyst consists of a combination of magnetite ($Fe_3O_4$), iron carbides, and amorphous carbon [17-19]. Recent work from this group has also proposed a role for a hydrocarbonaceous overlayer [16, 20].

A sample of $\alpha$-$Fe_2O_3$ pre-catalyst was prepared using the precipitation technique in a similar manner to that described previously [16]. However, on this occasion, a Mettler-Toledo chemical preparation system (LabMax batch reactor) was used to ensure careful control of pH during the precipitation process. The resulting $\alpha$-$Fe_2O_3$ exhibited a BET surface area of 70.8 $m^2\,g^{-1}$, which exceeds the value of a previously reported catalyst [16]; with the higher surface area thought to reflect a more controlled precipitation in this case. The catalyst was ground and sieved to a size fraction of 250 - 500 μm. Given the fixed bed reactor adopted for these experiments, it is important to avoid the presence of fine catalyst particles which could aggregate on reaction, 'plug' the reactor and give rise to a significant pressure drop across the catalyst bed. Two samples were prepared for INS analysis; a sample exposed to dehydration conditions to serve as a background and subsequently reacted under the CO hydrogenation reaction conditions. This procedure is intended to provide information on how the surface of the catalyst may have changed as a consequence of a test reaction (CO hydrogenation at ambient pressure and elevated temperature) that produces gaseous products without inducing actual Fischer-Tropsch activity [16]. The latter option is avoided as it would generate high molecular weight products, which would complicate analysis of the INS spectrum. CO hydrogenation over iron based catalysts has previously been shown to provide useful information on aspects of Fe/CO/$H_2$ surface chemistry that is inherently relevant to F-T catalysis [16, 20-21].

The background sample was prepared as follows. A stainless steel reactor was charged with 17.2 g of the $\alpha$-$Fe_2O_3$ pre-catalyst, which was located between two sections of quartz wool. This mass of catalyst is smaller than used on previous occasions [16] but is sufficient to fill an aluminium INS cell and provide a spectrum with reasonable signal-to-noise ratio with *ca.* 8 h measurement time on the MAPS spectrometer. Further, the lower mass of catalyst coupled with the superior gas handling facilities described in Section 2.1 means that weight hourly space velocities closer to those encountered in a micro-reactor are possible. The catalyst was purged by passing helium at a flow rate of 1000 ml $min^{-1}$ whilst heating to 373 K at a rate of 5 K $min^{-1}$, then maintaining that temperature for 12 h. This procedure desorbs any water that may be absorbed within the pore structure of the sample. After this treatment, the heating was stopped, the cell isolated and transferred to an argon-filled glove box (MBraun UniLab MB-20-G, [$H_2O$] < 1 ppm, [$O_2$] < 2 ppm), where the dried catalyst was transferred to a standard aluminium INS cell [15], which was sealed *via* an indium wire gasket. The sample could then be transported to the INS spectrometer.

The $\alpha$-$Fe_2O_3$ sample selected for CO hydrogenation was exposed to a feedstream comprising CO and $H_2$ in a 1:2 ratio (CO flow rate = 75 ml $min^{-1}$, $H_2$ flow rate = 150 ml $min^{-1}$) in the presence of a helium diluent (600 ml $min^{-1}$) at ambient pressure as the temperature was ramped up to 623 K at 5 K $min^{-1}$ then held at 623 K for 6 h. These conditions have previously been shown to support CO hydrogenation [16]. Figure 5 shows the (uncalibrated) mass spectrometer response as a function of time-on-stream with respect to the temperature programme. The profile for reagents and products broadly follows that recently reported for a comparably specified iron-based F-T catalyst [16]. An initial consumption of CO at *ca.* 600 K is associated with the conversion of hematite (α-$Fe_2O_3$) to magnetite ($Fe_3O_4$). A second more gradual consumption of CO is seen once the operating temperature of 623 K is reached, which then progressively diminishes and approaches a steady state value at about 7 h. Relatively minor inflections in the dihydrogen signal up to t = 1.5 h indicate that hematite reduction is primarily facilitated using CO rather than $H_2$ as the reductant. From t = 1.5 h onwards a small consumption of dihydrogen in the feedstream is connected with the CO profile over this period and is associated with methane formation. A gradual decline in methane production is associated with a slow deactivation channel.

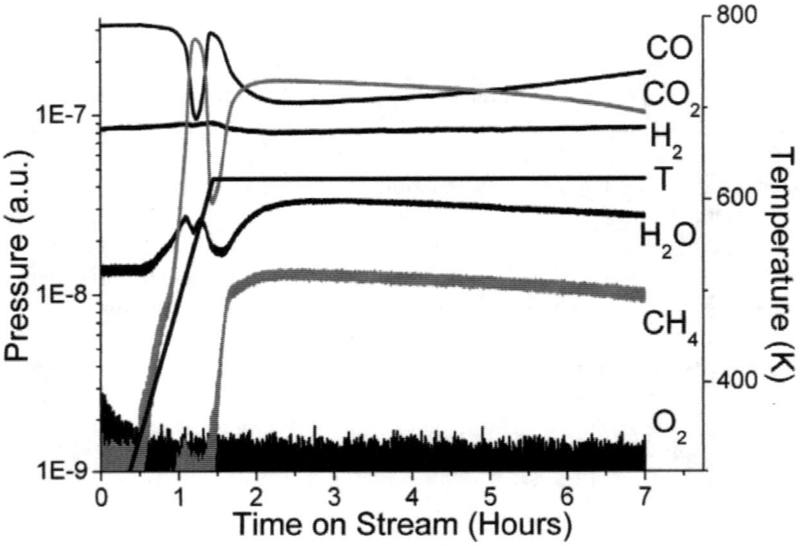


**Figure 5** Mass spectrometer response (uncalibrated) defining the reaction profile for the CO hydrogenation reaction at 623 K over the α-$Fe_2O_3$ sample which is located in the INS sample cell (Figure 3). The mass traces are labelled on the right hand side of the profile in descending order. T defines the temperature profile over the seven hour period.

The production of $CO_2$ and $H_2O$ during the temperature ramp could be the result of the stepwise reduction of the α-$Fe_2O_3$ starting material towards $Fe^0$ *via* $Fe_3O_4$ and FeO by CO and $H_2$ resulting in the evolution of $CO_2$ and $H_2O$ respectively [22]. Several contributing pathways could be responsible for the second $CO_2$ feature at *ca.* 623 K: CO disproportionation via the Boudouard reaction [23] and/or CO oxidation with surface oxygen. The increase in $H_2O$ signal at 623 K could be attributed to the hydrogenation of surface oxygen after CO dissociation. The water gas shift reaction [24] is also thought to be active in consuming CO at 623 K to form $CO_2$ and $H_2$. A more comprehensive discussion of reaction profiles associated with CO hydrogenation over iron based F-T catalysts is available elsewhere [16].

After 6 h on stream, the reaction was quenched by terminating the syngas flow and allowing the sample to cool to room temperature in flowing helium. As with the background sample, the cell was isolated and placed within the argon-filled glove box. The reacted catalyst sample was transferred to a standard aluminium INS cell and then transported to the INS spectrometer for spectral acquisition.

Journal of Physics: Conference Series **554** (2014) 012005 doi:10.1088/1742-6596/554/1/012005

### 3.2 Inelastic neutron scattering

The INS spectra of the dehydrated and reacted samples were recorded using the MARI spectrometer, which is located at the ISIS Facility of the STFC Rutherford Appleton Laboratory. A description of the operational characteristics of the MARI spectrometer is presented elsewhere [2,25]. Spectra were recorded at 20 K and at an incident neutron energy of 4840 $cm^{-1}$ using the A-chopper package. An acquisition period of 8 h was used for both the background and reacted spectra.

The spectrum (2000-4000 $cm^{-1}$) of the dehydrated $\alpha$-$Fe_2O_3$ before reaction is presented in Figure 6(a), whereas the catalyst post-reaction is presented in Figure 6(b). The unreacted $\alpha$-$Fe_2O_3$ spectrum exhibits no features in the 2000-4000 $cm^{-1}$ region; in direct contrast to a recently reported INS spectrum of an iron based F-T catalyst that was prepared in-house, where a distinct band at 3442 $cm^{-1}$, assigned to the O-H stretch of hydroxyl groups present as surface terminations of the hematite precursor compound, is observed [16]. The absence of a ν(O-H) band in Figure 6(a) indicates that the $\alpha$-$Fe_2O_3$ examined here differs from the previously studied in-house catalyst; possibly as a consequence of the tighter control of the precipitation stage used in the catalyst preparative part of this work that has prevented generation of hydroxyl groups at the catalyst surface. Often hydroxyl groups are present at metal oxide surfaces as a means of minimizing the chemical potential in the surface region, as the bulk metal oxide structure terminates. Direct geometry INS spectrometers are particularly good at characterizing hydroxyl groups associated with high surface area metal oxides [7] and, as hydroxyl groups can play a major role in influencing surface chemical reactions [7], this is an area where INS is expected to find increasing application.

The INS spectrum of the reacted catalyst (Figure 6(b)) is characterized by two distinct features. Firstly, a sharp band is observed at 3061 $cm^{-1}$, which is assigned to a C-H stretch of olefinic or aromatic C-H bonds ($sp^2$ hybridised C atoms) present at the catalyst surface. Secondly, a low frequency shoulder is evident on the main peak at 2942 $cm^{-1}$, which is assigned to an aliphatic ν(C-H) mode ($sp^3$ hybridised C atoms). Thus, Figure 6(b) indicates a population of predominantly $sp^2$-C–H species has formed at the catalyst surface upon reaction, which is supplemented by a lower density of aliphatic C-H species. Following recent INS investigations of iron based F-T catalysts [16, 20], Figure 6 is interpreted as indicating the presence of a hydrocarbonaceous overlayer. Traditionally, iron based F-T catalysts are typically analysed by techniques such as X-ray diffraction, Mössbauer spectroscopy, electron microscopy, *etc.* and, consequently, described in terms of solid-state chemistry (hematite, magnetite, iron carbides, *etc.*) [17-19], whilst techniques such as temperature-programmed oxidation and Raman scattering highlight the relevance of carbonaceous materials [26-27]. This new application of INS offers the exciting potential of being able to determine how hydrogen atoms are being partitioned within the F-T catalyst matrix. In order to better understand the surface chemistry connected with the lucrative associated process chemistry, a role for hydrogenous species needs to be more precisely defined. This combination of INS and an analysis of CO hydrogenation reactions over iron based F-T catalysts, representative of industrial grade catalysts [16, 20] is providing new knowledge that complements information accessible via the more traditional analytical methods.

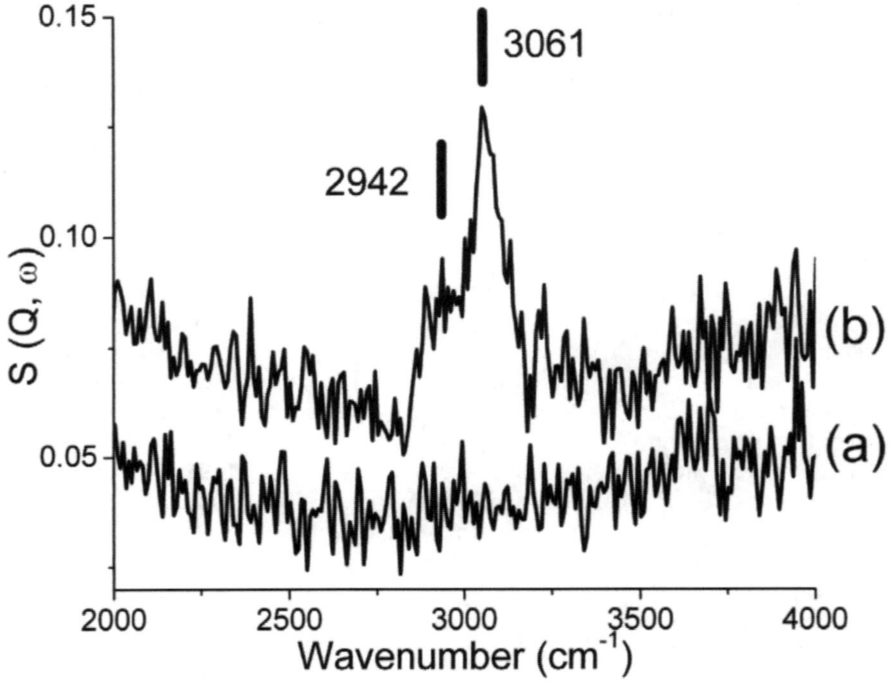


**Figure 6.** Inelastic neutron scattering spectra of (a) the dehydrated iron oxide catalyst before reaction and (b) the post-reaction catalyst sample after reaction at 623 K for 6 hours. Spectra were recorded at 20 K and an incident energy of 4840 $cm^{-1}$.

## 4. Conclusions

Recent improvements to sampling environment methodologies connected with the acquisition of INS spectra of heterogeneous catalysis samples at the ISIS Facility are described. This includes the re-configuration and up-grading of a sample preparation laboratory, the commissioning of a new gas handling apparatus and the construction of new reactors. Together, these facilities are able to support a broad programme of investigations of heterogeneous catalysts featuring chemistry at the gas/solid interface. As an illustration of the type of work presently utilizing these enhanced sample-handling facilities, the INS spectrum of an iron based F-T catalyst is presented before and after sustained CO hydrogenation at 623 K. The resulting spectra, recorded using the MARI spectrometer, provide further evidence for a role for a hydrocarbonaceous overlayer in F-T related chemistry.

## Acknowledgements

The laboratory refurbishment and construction of new reactors was funded by the STFC Rutherford Appleton Laboratory. The gas handling apparatus was funded with support from the STFC Rutherford Appleton Laboratory and the EPSRC (Grant No. EP/E028861). Sasol Technology UK Ltd. and the University of Glasgow are thanked for the provision of an Industrial Studentship (RW). The Erasmus scheme of the EU Lifelong Learning Programme is thanked for student support (DB).

## References

[1] Albers P W and Parker S F 2007 *Adv. Catal.,* **51** 99
[2] McInroy A R, Lundie D T, Winfield J M, Dudman C C, Jones P, Parker S F, Taylor J W and Lennon D 2005 *Phys. Chem. Chem. Phys.* **7** 3093
[3] Lundie D T, McInroy A R, Marshall R, Winfield J M, Jones P, Dudman C C, Parker S F, Mitchell C and Lennon D 2005 *J. Phys. Chem. B* **109** 11592
[4] Parker S F, Lennon D and Albers P W 2011 *Appl. Spectrosc.* **65** 1325
[5] Ryczkowski J 2001 *Catal. Today* **68** 263
[6] El-Roz M, Bazin P, Daturi M and Thibault-Starzyk F 2013 *ACS Catal.* **3** 2790
[7] McInroy A R, Lundie D T, Winfield J M, Dudman C C, Jones P, Parker S F and Lennon D 2006 *Catal. Today* **114** 403
[8] Silverwood I P, Hamilton N G, Staniforth J Z, Laycock C J, Parker S F, Ormerod R M and Lennon D 2010 *Catal. Today* **155** 319
[9] Silverwood I P, Hamilton N G, McFarlane A R, Kapitan J, Hecht L, Norris E L, Ormerod R M, Frost C D, Parker S F and Lennon D 2012 *Phys. Chem. Chem. Phys.* **14** 15214
[10] McFarlane A R, Silverwood I P, Warringham R, Norris E L, Ormerod R M, Frost C D, Parker S F and Lennon D 2013 *RSC Advances* **3** 16577
[11] McFarlane A R, Silverwood I P, Norris E L, Ormerod R M, Frost C D, Parker S F and Lennon D 2013 *Chem. Phys.* **427** 54
[12] Silverwood I P, Hamilton N G, McFarlane A, Ormerod R M, Guidi T, Bones J, Dudman M P, Goodway C M, Kibble M, Parker S F and Lennon D 2011 *Rev. Sci. Instrum.* **82**
[13] Chinta S, Choudhary T V, Daemen L L, Eckert J and Goodman D W 2002 *Angew. Chem. Int. Ed.* **41** 144
[14] Sivadinarayana C, Choudhary T V, Daemen L L, Eckert J and Goodman D W 2004 *J. Am. Chem. Soc.* **126** 38
[15] Mitchell P C H, Parker S F, Ramirez-Cuesta A J and Tomkinson J 2005 *Vibrational spectroscopy using neutrons.* (Singapore, World Scientific)
[16] Hamilton N G, Warringham R, Silverwood I P, Kapitan J, Hecht L, Webb P B, Tooze R P, Zhou W, Frost C D, Parker S F and Lennon D 2014 *J. Catal.* **312** 221
[17] Shroff M D, Kalakkad D S, Coulter K E, Kohler S D, Harrington M S, Jackson N B, Sault A G and Datye A K 1995 *J. Catal.* **156** 185
[18] Niemantsverdriet J W and Van der Kraan A M 1981 *J. Catal.* **72** 385
[19] Niemantsverdriet J W, Van der Kraan A M, Van Dijk W L and Van der Baan H S 1980 *J. Phys. Chem.* **84** 3363
[20] Hamilton N G, Silverwood I P, Warringham R, Kapitan J, Hecht L, Webb P B, Tooze R P, Parker S F and Lennon D 2013 *Angew. Chem. Int. Ed.* **52** 5608
[21] Graf B, Schulte H and Muhler M 2010 *J. Catal.* **276** 66
[22] Kock A, Fortuin H M and Geus J W 1985 *J. Catal.* **96** 261
[23] Kadkhodayan A and Brenner A 1989 *J. Catal.* **117** 311
[24] Dry M E 1982 *Chem. Tech.* **12** 744
[25] Arai M, Taylor A D, Bennington S M and Bowden Z.A 1992 *Recent developments in the physics of fluids*, ed. Howells W S and Soper A K (Bristol, Adam Hilger) F321–F328
[26] Herranz T, Rojas S, Perezalonso F, Ojeda M, Terreros P and Fierro J 2006 *J. Catal.* **243** 199
[27] Dictor R A and Bell A T 1986 *J. Catal.* **97** 121

Journal of Physics: Conference Series **554** (2014) 012006 doi:10.1088/1742-6596/554/1/012006

# Hydrogen gas sample environment for TOSCA

**Mark G Kibble, Anibal J Ramirez-Cuesta, Chris M Goodway, Beth E Evans and Oleg Kirichek**

ISIS Facility, Rutherford Appleton Laboratory, Chilton, Didcot, Oxfordshire, OX11 0QX, United Kingdom

E-mail: oleg.kirichek@stfc.ac.uk

**Abstract**. The idea of using hydrogen as a fuel has gained immense popularity over many years. Hydrogen is abundant, can be produced from renewable resources and is not a greenhouse gas. However development of hydrogen based technology is impossible without understanding of physical and chemical processes that involve hydrogen sometime in extreme conditions such as high pressure or low and high temperatures. Neutron spectroscopy allows measurement of a hydrogen atom motion in variety of samples. Here we describe and discuss a sample environment kit developed for hydrogen gas experiment in a broad range of pressure up to 7 kbar and temperatures from 4 K to 473 K. We also describe para-hydrogen rig which produces para-hydrogen gas required for studying the rotational line of molecular hydrogen.

## 1. Introduction

Research on a viable and safe storage medium for hydrogen gas is becoming increasingly important for the development of a clean hydrogen-based economy. Storage of hydrogen gas in solids, able to act as reversible sponges for hydrogen absorption and desorption, is in that sense one of the most promising areas of hydrogen storage medium development [1]. Neutron spectroscopy allows measurement of hydrogen atom motion in solid state samples, which can explain the growing popularity of this technique in hydrogen storage material research.

While neutron diffraction is used to determine and investigate the crystal structure of the material, spectroscopy can measure the atom dynamics and obtain full phonon dispersion curves without the limitations typical in other vibrational spectroscopy techniques (e.g., Raman and IR).

Spectroscopic studies by neutrons are traditionally very well suited to looking into hydrogen physisorption phenomena. Most of the studies that have been done so far using the TOSCA spectrometer [2] refer to low temperature and high pressure experiments. But, thanks to recent instrument upgrades and improvements to the ISIS neutron source, a whole new range of possibilities has opened: even hydrogen chemisorption events can be investigated, requiring much higher temperatures than the conventional use at 77 K. Also hydrogen absorption in nano-porous materials at high pressure could be effectively investigated by TOSCA measurements, but has not been done before mainly due to the lack of sample environment equipment capable of reaching the required temperatures and pressures.

In this paper we describe a sample environment kit developed for hydrogen gas experiments on TOSCA instrument which consists of para-hydrogen gas handling rig, gas handling panels, high pressure intensifiers, sample sticks and sample cells. The kit provides sample environment in a broad range of pressures from vacuum up to 7 kbar and temperatures 4 K – 473 K. The para-hydrogen rig produces para-hydrogen gas required for studying the rotational line of molecular hydrogen. This sample environment kit allows the investigation of hydrogen properties in confinement from a fundamental as well as from an applied point of view, giving full insight into the dynamics of molecular hydrogen in relation to the host matrix. The kit can be also used on other neutron scattering instruments. Experimental results that demonstrate the use of the para-hydrogen rig in TOSCA experiments are also presented.

## 2. Para-hydrogen gas handling rig

Normal hydrogen gas contains 25% para-hydrogen and 75% ortho-hydrogen. These two forms of hydrogen have different properties and are very useful to researchers studying hydrogen, particularly to those using neutron scattering. Normal hydrogen may be converted to almost 100% para-hydrogen at 20 K [3]. In the case of isolated or non-interacting molecules conversion is usually prevented. However a magnetic field gradient will catalyse this conversion. In the liquid or solid phase the conversion to para-hydrogen occurs very slowly, at a rate of less than 2% conversion per hour, via the interaction of magnetic dipole moments. In the low-density gas phase the conversion rate is extremely slow and high purity para-hydrogen samples can be kept for up to a year with very little reversion. There are techniques that allow the preparation of ortho and para-hydrogen with more than 99% purity.

In the case of liquid hydrogen the natural conversion to a 99% level of para-hydrogen purity is obtained after a few days. If the liquid is put in contact with a paramagnetic catalyst ($CrO_3$) the conversion can happen within minutes.

Formerly, para-hydrogen was produced at ISIS using a conventional liquid cryogen-based cryostat. The operation of this kind of system required significant resources, created a number of logistical issues including the considerable cost of the required cryogens and posed health and safety problems. The progress in Closed Cycle Refrigerator (CCR) technology now offers a new generation of cryogenic systems with significantly reduced consumption and in some cases the complete elimination of cryogens [4]. There are two fundamentally different CCR based approaches. First, so-called "dry" systems that do not contain liquid cryogens at all, built around a CCR that utilises the cooling power produced by the cold head [5, 6]. The second option is based on the idea of re-condensing the evaporating helium back to a standard cryostat by a CCR [7]. The ISIS sample environment group has decided to design and construct a specific cryogen-free para-hydrogen rig based on the "dry" option.

The cryogenic part of the para-hydrogen rig is shown in figure 1. The assembly consists of a para-hydrogen generating cell, containing a catalyst, mounted on a CCR (Leybold 1040 cold head CR-05 with base temperature of 10K) and wired with temperature sensor and heater. The cell is connected to a gas handling system through 1/8" capillary tube for evacuation and for the control of the hydrogen supply and para-hydrogen extraction.

The base temperature of the para-hydrogen generating cell is 16 K, obtained in 4 hours from room temperature. The vibrations produced by the mechanical cooler sufficiently mix the hydrogen to bring about the maximum possible conversion rate through contact with the catalyst. The conversion process takes less than 12 hours for an efficiency of 98% conversion. Neutron measurements on gas samples have confirmed the maximum conversion to para-hydrogen, but the quality of the para-hydrogen may also be verified in-situ with a custom-built gauge. This gauge uses the difference in thermal conductivities of normal and para-hydrogen to determine the relative percentages of the two forms of hydrogen in a gas sample. Using this gauge the para-hydrogen percentage can be determined with $<$ 2% error.

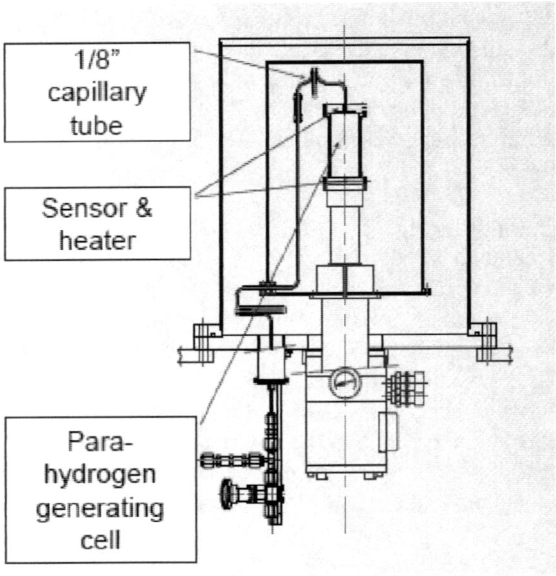


**Figure 1.** TOSCA para-hydrogen generating rig.

## 3. High pressure hydrogen gas sample environment

The ISIS sample environment group has a broad range of equipment for enabling experiments using high pressure inert gas cells [8, 9]. However the development of components for high pressure hydrogen systems is more difficult because of hydrogen embrittlement: a process whereby certain materials become brittle following exposure to hydrogen. High-strength steels, titanium and aluminium alloys seem particularly vulnerable to this form of corrosion, and fracture can often result from the ingress of atomic hydrogen into a material's crystal structure. During the development of equipment for high pressure hydrogen experiments on the TOSCA spectrometer special attention was paid to the hydrogen compatibility of materials and safety aspects of the equipment design.

The TOSCA high pressure hydrogen kit consists of gas handling panels, high pressure intensifiers, sample sticks and sample cells. The gas handling panels can be used for gas dosing (known volumes), gas mixing, gas flow (including moisture carrying), gas absorption/desorption and condensing, volumetric measurement and catalyst work. All the panels are fitted with pressure relief devices. The standard hydrogen gas handling panel covers a working pressure (WP) range from vacuum to 200 bar. The panel also contains a 124 bar WP 500cc buffer, three gas supply feeds, a pressure transducer and baratron.

## 4. High pressure hydrogen gas intensifier

For high pressure hydrogen gas experiments up to 3.5 kbar a two stage intensifier is used which consists of a first stage diaphragm pump (up to 2500 bar), a second stage hand intensifier (up to 3500 bar), transducer readouts that allow 1bar control increments (via the hand intensifier) and a safety burst disc protection system. The hydrogen two stage intensifier has been used in an experiment to maintain a pressure of 2800 bar for a five day period in an experimental cell at a temperature of 20 K. This experiment was conducted using the Inconel high pressure cell RLI 469, which is one of the two hydrogen high pressure Inconel cells within the ISIS sample environment group.

As a part of the Joint Research Activities FP7 [9] project, ISIS committed to building a 10 kbar hydrogen intensifier. Due to financial restraints, and the difficulty in sourcing a suitable system, it was decided to assemble an intensifier from commercially available components at ISIS. Hydrogen Intensifier with similar technical specifications has been developed at Hahn Meitner Institute (now

Helmholtz-Zentrum Berlin) [10]. This system uses different basic operational principle: the intensifier case is sealed and filled with an inert gas. Sensitive hydrogen gas sensors automatically shut the system down as soon as any presence of hydrogen is detected in the case. As a result the system has got almost zero tolerance to hydrogen gas leaks. The design of ISIS Intensifier allows small leaks which significantly relax demand for the system leak tightness. The system controller constantly monitors the pressure change rate in the high pressure part of the intensifier and if the rate exceeds established limit the system shuts itself down in a safe way. The intensifier frame is a fully welded construction that contains front and rear removable cover panels for accessing the internal components. The frame also includes an electrically isolated and hermetically sealed compartment for electronics. One cover panel houses the extraction port connection for intense ventilation of the system with nitrogen or other inert gas. The intensifier can be transported around the experimental hall using a lifting frame and then wheeled into position by the instrument. The 10 kbar hydrogen intensifier is similar in design to a standard gas intensifier, but with semi-automatic control of the high pressure stage.

The intensifier has a 'Burst disc recovery system' fitted to safely contain $H_2$ vented in the event of any burst disc rupture in the system including the $H_2$ centrestick. The 'Burst disc recovery system' comprises of 3 linked 3875cc buffer volumes (1bar Argon filled) connected to the intensifier system burst discs and centrestick burst disc within a closed circuit fitted with monitoring a pressure transducer and pressure relief valve.

The intensifier has been assembled and tested with helium gas up to 10 kbar and hydrogen gas up to 8.5 kbar. The further testing of the system with hydrogen gas is in progress.

## 5. High pressure centre-sticks

The ISIS hydrogen gas 10 kbar centre-stick allows a sample cell to be mounted to enable gas loading. The pressure and temperature can be monitored and the system is protected via a pressure relief device. The centre-stick can be used with 100 mm diameter variable temperature inserts and covers a pressure range from vacuum to 10 kbar of working pressure. A 3/16" heated capillary, 10 kbar Stansted transducer, sample cell thermometry and heating connections are also integrated in the design of the centre-stick.

## 6. High pressure sample cells

Tosca high pressure sample cells are manufactured from neutron friendly materials with high resistance to hydrogen gas exposure to compliment the TOSCA instrument and experimental requirements. The cell wall thickness is a compromise between minimising the cell neutron scattering background and providing enough strength for the safe operation of the cell at its specified pressure and temperature.

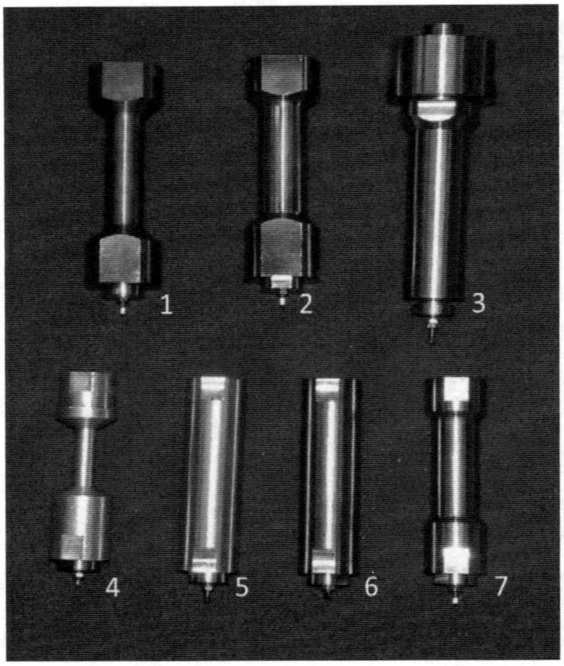


**Figure 2.** High pressure TOSCA sample cells; (1) 6 kbar hydrogen gas BeCu cell with temperature up to 20°C; (2) 7.36 kbar hydrogen gas BeCu cell up to 20°C; (3) 8 kbar hydrogen gas BeCu/TiZr cell (sleeved cell) up to 20°C; (4) 300 bar hydrogen gas Aluminium cell up to 100°C; (5) 4.4 kbar hydrogen gas Aluminium cell up to 20°C; (6) 5.4 kbar inert gas TiZr cell up to 20°C; (7) 6.35 kbar hydrogen gas Inconel cell up to 20°C.

The sample cells presented in figure 2 can be used in a passive or pressure rated regime for powder, liquid or solid samples. There are seven high pressure cells which can be used in TOSCA experiments: (1) 6 kbar hydrogen gas BeCu cell for temperatures up to 20°C, volume 2.23cc; (2) 7.36 kbar hydrogen gas BeCu cell for temperatures up to 20°C, volume 2.25cc; (3) 8 kbar hydrogen gas BeCu/TiZr cell (sleeved cell) for temperatures up to 20°C, volume 2.8cc; (4) 300 bar hydrogen gas Aluminium (7075 T6 Alloy) cell up to 100°C, volume 1.76cc; (5) 4.4 kbar hydrogen gas Aluminium (7075 T6 Alloy) cell up to 20°C, volume 1.77cc; (6) 5.4 kbar inert gas TiZr cell up to 20°C, volume 1.77cc; (7) 6.35 kbar hydrogen gas Inconel cell up to 20°C volume 1.77cc. All materials used for cells are known for high resistance to hydrogen gas exposure [11]. However the embrittlement is resulting from different chemical reactions in copper contained alloys [12] and steel [13] which can explain why BeCu is more popular as a material for hydrogen gas pressure cells. All the cells are tested and registered according to the ISIS high pressure vessel code of practice and have a limitation on the number of experiments they can be used in. For example the design service life of cell (1) is 400 cycles with periodical inspection at (1/2 design life) 200 cycles.

## 7. Neutron spectroscopy experimental results

An example of the use of high pressure cells to study the rotational dynamics of solid para-hydrogen is shown in figure 3.

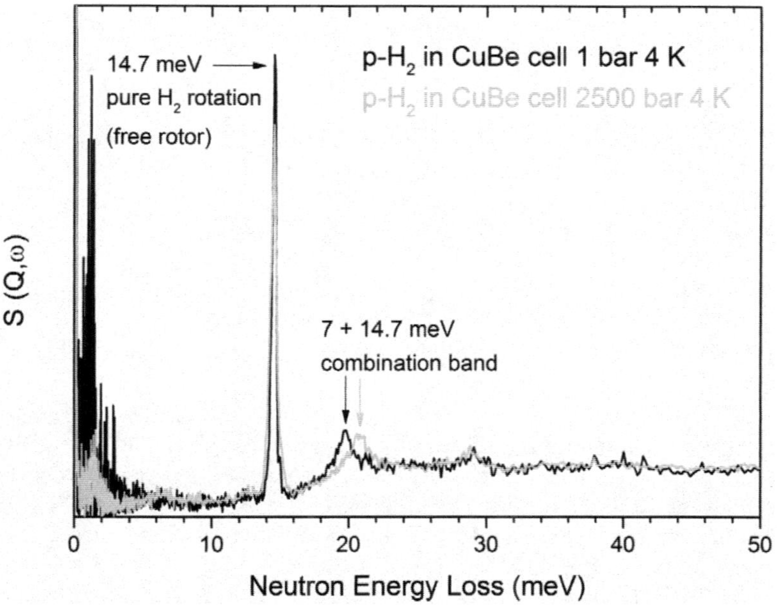


**Figure 3.** Inelastic neutron scattering spectra of solid para-hydrogen (p-$H_2$) recorded at a temperature of 4 K for two different pressures: 1 bar (black trace) and 2500 bars (green trace). Namely in both cases the solid p-$H_2$ was kept under pressure of normal $H_2$.

Para-hydrogen was produced with the para-hydrogen rig described above. The cell is a BeCu high pressure cell shown in figure 2. The cell is cooled below 20 K and para-hydrogen is dosed into the cell. The cell is then pressurised to 5 bar with para-hydrogen and the temperature dropped to 4 K. Once the cell is cold, the centre-stick is detached from the para-hydrogen rig and connected to the high pressure rig so that the high pressure normal hydrogen can be added to the system. The signal observed is mostly para-hydrogen with the little contamination that arises from the introduction of normal hydrogen at high pressure. In order to minimise multiple scattering during the experiment the cell is filled with a solid cylinder made from aluminium, leaving an annular space of 0.1 mm. The inelastic neutron scattering spectra of solid para-hydrogen (p-$H_2$) shown in figure 3 was recorded at a temperature of 4 K for two different pressures: 1 bar (black trace) and 2500 bars (green trace). As the normal hydrogen pressure increases the solid para-hydrogen stiffens and consequently the combination band which corresponds to the simultaneous excitation of rotational (14.7 meV) and lattice vibrational modes (7 meV) [14] is somewhat shifted toward higher wavenumbers.

## 8. Conclusions

Here we presented recently developed sample environment kit for hydrogen gas experiment in a broad range of pressure up to 7 kbar and temperatures from 4 K to 473 K. We also describe para-hydrogen rig which produces para-hydrogen gas required for studying the rotational line of molecular hydrogen. This sample environment kit allows the investigation of hydrogen properties in confinement such as hydrogen absorption in nano-porous materials or hydrogen chemisorption.

**Acknowledgements**
We are grateful to Prof Felix Fernandez-Alonso, and Drs Stewart F. Parker and Svemir Rudic for valuable discussions, help with data analysis and comments about the manuscript. This research project has been supported by funding under the STFC Facility Research and Development Scheme and by the European Commission under the 7th Framework Program JRA (Joint Research Activity)

through the 'Research Infrastructures' action of the 'Capacities' Programme, Contract No: CP-CSA_INFRA-2008-1.1.1 Number 226507-NMI3.

**References**

[1] Grochala W and Edwards P P 2004 Thermal decomposition of the non-interstitial hydrides for the storage and production of hydrogen *Chem. Rev.* 104 1283-1315

[2] Colognesi D, Celli M, Cilloco E, Newport R J, Parker S F, Rossi-Albertini V, Sacchetti F, Tomkinson J and Zoppi M 2002 TOSCA neutron spectrometer: the final configuration *Appl. Phys. A* 74 S64

[3] Silvera I F 1980 The solid molecular hydrogens in the condensed phase: fundamentals and statistical properties *Reviews of Modern Physics* **52**, 393-452

[4] Kirichek O 2012 Impact of the cryogen free revolution on neutron scattering laboratories *Modern Physics Letters B* **26** 1230006

[5] Oliver E C, Evans B E, Chowdhury M A H, Major R A, Kirichek O and Bowden Z A 2008 Novel testing chamber for neutron scattering measurements of internal stresses in engineering materials at cryogenic temperatures *Meas. Sci. Technol.* **19** 034019

[6] Evans B E, Down R B E, Keeping J, Kirichek O and Bowden Z A 2008 Cryogen-free low temperature sample environment for neutron scattering based on pulsed tube refrigeration *Meas. Sci. Technol.* **19** 034018

[7] Kirichek O, Carr P, Johnson C and Atrey M 2005 Nuclear magnetic resonance magnet actively cooled by pulse tube refrigerator *Rev. Sci. Instrum.* **76** 055104

[8] Done R, Kirichek O, Evans B E and Bowden Z A 2010 NMI6, FP7, JRA: Report on current inert gas pressure cell technology *arXiv* 1007.3135

[9] Kirichek O, Done R, Goodway C M, Kibble M G, Evans B E and Bowden Z A 2012 Development of high pressure gas cells at ISIS *Journal of Physics: Conference Series* **340,** 012008

[10] Michael Meißner and Dirk Wallacher private communication

[11] Louthan M R, Caskey G R, Donovan J A and Rawl D E 1972 Hydrogen embrittlement of metals *Materials Science and Engineering* **10,** 357

[12] Fast V D 1965 Interaction of mettals and gases *Academic Press*, New York, 54

[13] Weiner L C 1961 *Corrosion* **17**, 109

[14] Fernandez-Alonso F, Bermejo F J and Saboungi M L 2011 Molecular Hydrogen in Carbon Nanostructures *Handbook of Nanophysics: Functional Nanomaterials*, ed K D Sattler (Boca Raton, FL, US: CRC Press) chapter **40** 40

# Cryogenic sample environment on TOSCA

**Richard B E Down, Anibal J Ramirez-Cuesta, Robert A Major, Jeff Keeping, Svemir Rudić and Oleg Kirichek**

ISIS Facility, Rutherford Appleton Laboratory, Chilton, Didcot, Oxfordshire, OX11 0QX, United Kingdom

E-mail: oleg.kirichek@stfc.ac.uk

**Abstract.** We have developed a new top-loading cryogenic insert that forms part of the integrated cryogenic sample environment for the Tosca spectrometer at the ISIS Facility. New advanced cryo-coolers and optimised design of the cryogenic insert allowed us to dramatically decrease system cool down time from 20 to 7 hours, sample cooling time from 3 hours to 45 minutes and also reduce the base temperature of the insert from 12 K to 4.5 K in the continuous regime and ~1.5 K in the single shot regime. We have also reduced the number of cold heads from three to two. This lets us save the significant cost of the third cold head maintenance and has cut electricity consumption by ~30%. A much lower base temperature of 4.5 K has opened new opportunities for molecular spectroscopy studies on Tosca.

## 1. Introduction

Tosca is an indirect geometry spectrometer optimised for the study of molecular vibrations in the solid state [1]. Science on Tosca includes studies of catalysts, hydrogen storage materials, hydrogen bonded systems, advanced materials, biological samples and organic compounds. For the majority of experiments Tosca requires cryogenic sample environment. This is why the cryogenic sample chamber was embedded in the Tosca instrument. The design of the original sample chamber has been based on a dry cryogen free approach [2]. Necessary cooling power has been provided by three closed cycle refrigerators (CCR), which allowed measurements in the range between 12 K and room temperature. However the system suffered from a very slow initial cool-down of more than 20 h and time consuming sample changes and subsequent cooling of around 3 h.

The ISIS cryogenic section together with the Tosca instrument scientists have re-designed the cryogenic chamber in order to optimise the cryogenic performance of the system. Due to much higher efficiency of the newly designed cryostat the number of cryo-coolers has been reduced from three to two, which has resulted in a 30% reduction in electricity consumption. The newly designed cryostat has also demonstrated a much lower base temperature of 4.5 K and significantly reduced system and sample cooling times.

Here we present the design of the optimised cryogenic sample chamber and discuss the system's performance in a proof of principle measurement.

## 2. Design of cryogenic sample changer

There are two entirely different cryogen free approaches: one is based on re-condensing of the evaporating helium back to the liquid cryogen based cryostat [3] and another one, is the so-called "dry" cryogenics that does not contain liquid cryogens at all. The last system is constructed around a closed cycle refrigerator (CCR) and utilises the cooling power produced by the cold head [4, 5]. The Tosca cryogenic sample chamber is a dry cryogen free system. The main advantage of this approach is the absence of liquid cryogens and all of the following consequences: no liquid cryogens top-ups, elimination of asphyxiation risk and a relative independence of helium supply troubles. However there is one significant drawback - the limited cooling power of the CCR. The most powerful Gifford-McMahon (GM) CCRs achieve no more than 1.5 W of cooling power at 4.2 K. In a conventional liquid helium based system, one can easily achieve more than an order of magnitude higher cooling power than in a dry system just by opening a needle-valve and allowing higher flow of liquid helium. The main consequence of this is the long cool-down time of the dry system. However this crucial parameter can be significantly improved by optimisation of the cryogenic design.

The Tosca cryostat was originally designed and manufactured by AS Scientific Ltd and cooling was supplied by three Leybold CCRs. Two Leybold RGD1245 cold heads were used to cool the main copper tube thermally linked to the sample space and a single stage Leybold RGS120 cold head was used to cool the outer radiation shield. The three cold heads gave the cryostat the name "Triplex". The Triplex system cooled from room temperature to a base temperature of 20 K in approximately 20 hours and had a sample change time of approximately 3 hours.

In the first phase of the Triplex upgrade project the two Leybold 1245 10K cold heads were replaced by two Sumitomo SRDK415 4 K cold heads. In order to reduce the sample change time a new capillary for transferring liquid helium into the bottom of the sample space was also added. The new 4 K cold heads have high cooling power on the first stage which enabled the redundancy of the third single stage CCR. This provided a large saving in running and maintenance costs. The removal of this cold head led to a change of name to "Duplex". The results of the upgrade lead to an overall improvement with the Duplex achieving a base temperature of 8 K. However the cool-down time was not as good as expected.

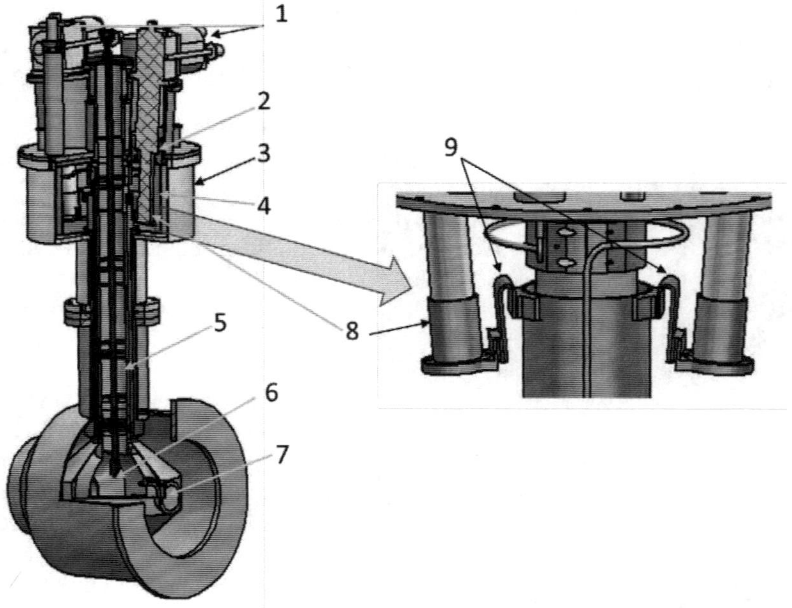


**Figure 1.** The new design of the Tosca cryostat: (1) two CCRs; (2) first stages of CCRs; (3) outer vacuum can; (4) infrared radiation shield; (5) cryogenic insert; (6) sample; (7) sample chamber with windows for neutron beam access; (8) CCR's second stages; (9) shimmed thermal links.

In order to improve the cool-down time it was decided to implement significant changes to the low-temperature part of the cryostat design. The new design presented in figure 1 was based on a modular approach in order to simplify any further repairs or modifications that could be required in the future. The cryogenic insert (5) is surrounded by an infrared radiation shield (4) attached to the first stages (2). The CCRs (1) are placed in the outer vacuum vessel (3). The thermal contact between the cold walls of the insert and the sample (6) is achieved by a small amount of heat-exchange helium gas added into the insert. The pressure of the heat-exchange helium gas is usually around 20 mbar.

The cooling power of CCR's second stages (8) is delivered to the copper tube of the insert through the shimmed thermal links (9). During preliminary tests of the newly designed cryostat significant temperature gradients were observed across the thermal links which pointed out that there was a "bottle-neck" in the thermal conductivity circuit between the CCR's second stages and the sample chamber (7).

In order to optimise the thermal conductance of the cryostat components, tests were carried out to compare the efficiency of different thermal linking techniques. Three copper thermal links were manufactured using different fixing techniques presented in figure 2: (A) clamping, (B) clamping with soldering and (C) welding. Each link was then mounted to the bottom of a 10K cold head with a dummy load fixed to the end of the link. Two temperature sensors were mounted to the links. The first was to read the temperature at the bottom of the cold head and the second was to read the temperature of the dummy load. Each thermal link was then cooled to its base temperature and the results of the three tests compared.

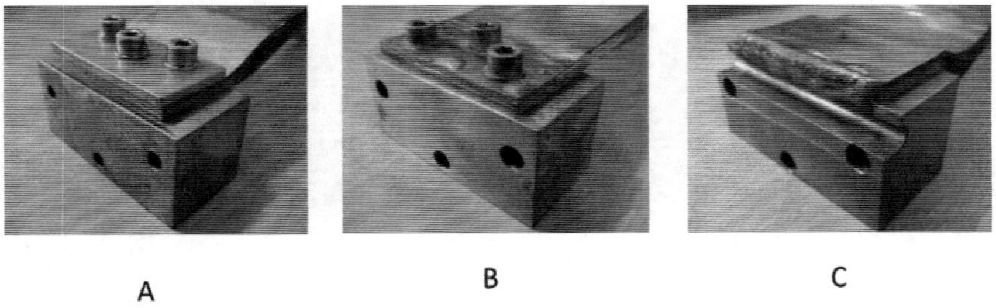


**Figure 2.** Three copper thermal links manufactured using different fixing techniques: (A) clamping; (B) clamping with soldering and (C) copper eutectic welding.

The data clearly showed the difference between the three fixing techniques. Initially clamping and clamping with solder were almost matched with clamping with solder cooling slightly faster. However the soldered links cooling began to slow around 50 K and only reached a base temperature of 33 K in 5 hours. The clamped link continued to cool to a base of 16 K in 4 hours. The copper eutectic welded links initial cooling was also similar yet maintained an almost linear cooling rate until it reached its base temperature of 14 K within 4 hours. With this new information it was decided to use the welded thermal links in the design of the cryostat.

## 3. System test results and discussion

The system cool-down test results are shown in figure 3a. It is possible to see the temperature of the sample chamber (1) lagging behind the temperature of the second stages of the CCRs (2). In the test the temperatures of both second stages were the same within our temperature resolution. After approximately three hours the temperature of the sample chamber started to catch up with the temperature of CCR's second stage and they became comparable after ~ 6.5 hours. The cryostat reached the base temperature of 4.5 K from start up in 7 hours, which demonstrates significant improvement in comparison with the previously achieved 20 hours of cool-down time to a base temperature of 12 K.

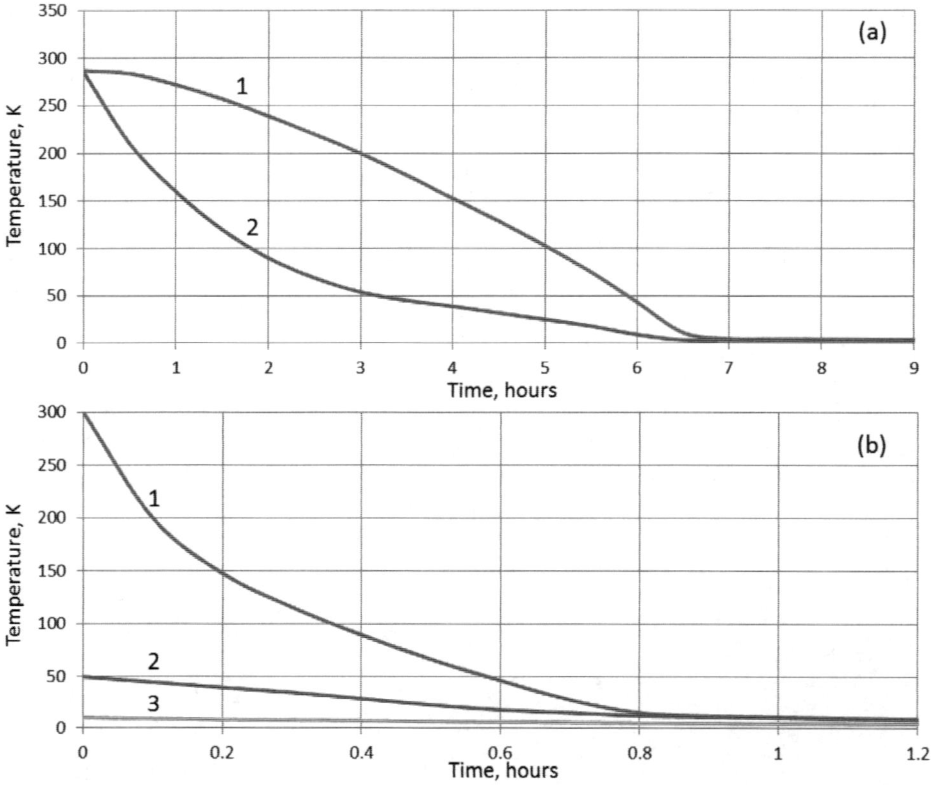


**Figure 3.** (a) The system cool down test results: (1) Temperature of the sample chamber; (2) temperature of the second stages of CCRs. (b) Cooling the sample from room to the base temperature with the cryogenic insert already cold: (1) temperature of the sample; (2) temperature of the sample chamber; (3) temperature of second stage of CCR.

In figure 3b we present the cool-down of the sample from room to the base temperature with the cryogenic insert already cold. As it is possible to see, the sample cooling from 300 K to 20 K takes just 45 minutes in comparison with previously achieved 3 hours. If required the sample can be rapidly cooled using a flow of liquid helium through the 4 mm helium transfer tube, that enters the cryostat at the sample space. In this case 4.5 K can be reached in less than 10 minutes. Pumping on the residual liquid helium remaining after completion of the sample cool down gives a temperature of ~ 1.5 K for approximately 3 hours in a single shot regime. However liquid helium assisted cooling turned out to be quite an expensive and inconvenient method, which can be explained by the high helium prices and the necessity of using trained personnel to perform this procedure. These disadvantages outweigh the short sample cooling time and lower base temperature. Therefore the use of liquid helium assisted cooling was restricted to a small number of experiments where a temperature of less than 2 K was essential.

Inelastic neutron scattering spectra of solid parahydrogen (p-$H_2$) recorded at a temperature of 4 K (black trace) and solid p-$H_2$ at 26 K (red trace) are presented in figure 4. In both cases the solid p-$H_2$ was kept under 2500 bar of normal $H_2$ pressure. The peak at 14.7 meV (117 $cm^{-1}$; 1 meV = 8.066 $cm^{-1}$) corresponds to the $H_2$ free rotor rotation. This transition is optically forbidden because in involves a nuclear-spin flip between ground-state p-$H_2$ (J = 0) and rotationally excited orthohydrogen, o-$H_2$ (J=1).

Furthermore, features at 7 and 20 meV are fundamental and combination bands which correspond to the most intense lattice vibrational frequencies in the solid and simultaneous excitation of rotational and lattice vibrational modes, respectively [6]. As the temperature is increased to 26 K the percentage of p-$H_2$ is reduced; consequently 14.7 meV peak as well as combination band are lower in intensity.

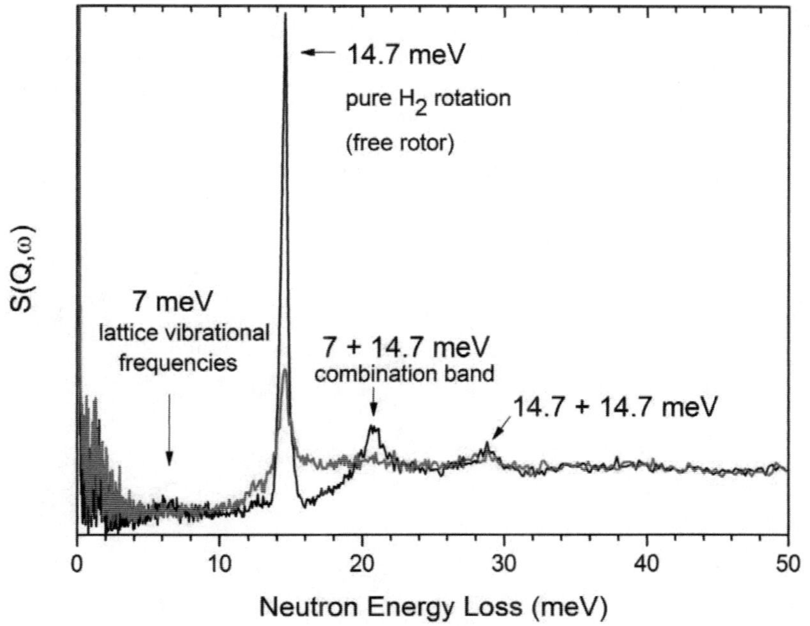


**Figure 4.** Inelastic neutron scattering spectra of solid p-$H_2$ recorded at temperature of 4 K (black trace) and solid p-$H_2$ at 26 K (red trace). At both temperatures the solid p-$H_2$ was kept under 2500 bar of normal $H_2$ pressure.

## 4. Conclusions

We have developed a new cryogenic insert for the Tosca inelastic neutron scattering spectrometer. New CCRs and optimised design of the cryogenic insert have allowed us to dramatically decrease the initial system cool down time to 7 hours, sample cooling time to 45 minutes and also to reduce the base temperature of the insert to 4.5 K in the continuous regime and 1.5 K in a single shot regime. We have also reduced the number of CCRs from three to two. This enables us to save the significant cost of the third CCR's maintenance and to cut electricity consumption by ~30%. The base temperature of 4.5 K has opened new realms of low temperature study on Tosca.

## Acknowledgements

We are grateful to members of the ISIS Experimental Operation Division involved in the Tosca cryogenics upgrade project. We would like to thank Prof Felix Fernandez-Alonso and Dr Stewart F. Parker for their help with data analysis and comments about the manuscript, as well as the TOSCA user community for their advice.

## References

[1] Colognesi D, Celli M, Cilloco E, Newport R J, Parker S F, Rossi-Albertini V, Sacchetti F, Tomkinson J and Zoppi M 2002 TOSCA neutron spectrometer: the final configuration *Appl. Phys. A* **74** S64

[2] Kirichek O 2012 Impact of the cryogen free revolution on neutron scattering laboratories *Modern Physics Letters B* **26** 1230006

[3] Kirichek O, Carr P, Johnson C and Atrey M 2005 Nuclear magnetic resonance magnet actively cooled by pulse tube refrigerator 2005 *Review of Scientific Instruments* **76** 055104

[4] Oliver E C, Evans B E, Chowdhury M A H, Major R A, Kirichek O and Bowden Z A 2008 Stress rig for neutron scattering measurements of bulk stress in engineering components at cryogenic temperatures *Meas. Sci. Technol.* **19** 034019

[5] Chapman C R, Evans B E, Dudman M P, Keeping J, Down R B E, Kirichek O and Bowden Z A 2011 Cryogen-free cryostat for neutron scattering sample environment. *Cryogenics* **51** 146

[6] Fernandez-Alonso F, Bermejo F J and Saboungi M L 2011 Molecular Hydrogen in Carbon Nanostructures *Handbook of Nanophysics: Functional Nanomaterials*, ed K D Sattler (Boca Raton, FL, US: CRC Press) chapter 40 pp 40-1

# Vibrational spectra of ferrocene, ferrocene-containing polymers and their oxidized compounds

**Markus Appel[1,2], Bernhard Frick[1], Alexandre Ivanov[1], Johannes Elbert[3], Matthias Rehahn[3], Markus Gallei[3], Tinka Luise Spehr[2] and Bernd Stühn[2]**

[1] Institut Laue-Langevin, 71 avenue des Martyrs, 38000 Grenoble, France
[2] Institute of Condensed Matter Physics, Technische Universität Darmstadt, Hochschulstraße 8, 64289 Darmstadt, Germany
[3] Ernst-Berl Institute for Chemical Engineering and Macromolecular Science, Technische Universität Darmstadt, Alarich-Weiss Straße 4, 64287 Darmstadt, Germany

E-mail: appel@ill.eu

**Abstract.** Ferrocene-containing polymers are an important member in the class of redox-responsive polymers, changing e.g. their solubility or conformation upon external stimuli. We present a study of their vibrational spectra using inelastic neutron scattering, focusing on the central building block of these polymers, the organometallic ferrocene complex. The vibrational modes of a bulk ferrocene sample are compared to those of poly(vinylferrocene), poly(ferrocenyldimethylsilane) and poly(ferrocenylmethylsilane). In the former polymer, the ferrocene complex is laterally attached to the polymer chain and the vibrational spectrum shows a slight shift and broadening of the fingerprint modes in the range of 100–800 $cm^{-1}$, except for the ring-metal-ring stretching mode which was not detectable anymore. The latter two polymers, where the ferrocene complex is part of the polymer backbone, exhibit larger differences to the vibrational spectrum of bulk ferrocene. Moreover, several contributions to the spectra caused by methyl groups in these polymers could be identified. In order to study the influence of oxidation on ferrocene and redox-responsive polymers, we investigated the ionic compound ferrocenium triiodide and oxidized poly(vinylferrocene). It is observed that the weakening of the $\eta^5$-complex bond by a missing electron leads to a significant shift of the fingerprint modes to lower frequencies.

## 1. Introduction

In the last decade stimuli-responsive polymers, which are able to alter their solubility, conformation or form/break covalent bonds upon an external trigger have attracted enormous attention [1–4]. Beside the vast majority of reports on temperature, light or pH responsive materials, other external triggers are also of growing interest [2, 5]. One of these less studied triggers is the redox-stimulus [6, 7]. Redox-responsive polymers can be produced, for example, by incorporating redox active organometallic sandwich complexes in the polymer [8, 9]. The most widely used organometallic building block is ferrocene, due to the excellent combination of redox, mechanical, semi-conductive, photo-physical and opto-electronic properties in the polymers formed [10–12]. For example, these polymers have been used in composite colloidal crystal films [13], self-healing materials [14] for redox-responsive release from nanocapsules [15], for switching surface wettability [16], for modulating the activity of catalysts [17] or for

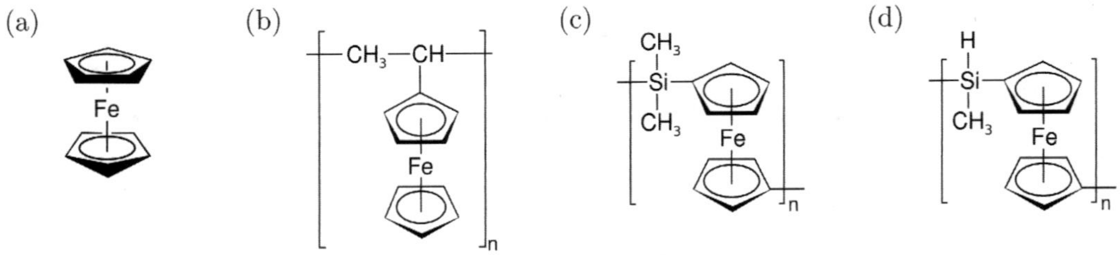


**Figure 1.** Chemical structure of (a): the ferrocene (Fc) molecule, and the ferrocene-containing polymers (b): poly(vinylferrocene) (PVFc), (c): poly(ferrocenyldimethylsilane) (PFDMS), and (d): poly(ferrocenylmethylsilane) (PFMS).

ion-selective membrane-gating [18]. In addition a vast amount of block copolymers with the ferrocene in the main chain, e.g., poly(ferrocenylsilanes), or with laterally bound ferrocene, e.g., poly(vinylferrocene), have been synthesized and their self-assembly in bulk and solution has been studied [10, 19–28].

Although many applications of these polymers have been explored, fundamental understanding of the local dynamics of ferrocene in these polymers and how it is influenced by oxidation is rather limited. With our experiments presented here, we aim to investigate the vibrational properties of the ferrocene moieties when incorporated in the macromolecular structure of polymers and/or being subject to oxidation. Vibrational spectroscopy using inelastic neutron scattering (INS) on ferrocene has been proven useful in previous studies of the interactions of isolated ferrocene molecules in supercages of zeolite [29, 30]. We now apply INS to compare the vibrational modes in bulk ferrocene, shown in figure 1(a), to different ferrocene-containing polymers. In poly(vinylferrocene) (PVFc), shown in figure 1(b), the ferrocene moiety is laterally attached to the polymer chain as side group, where poly(ferrocenyl-dimethylsilane) (PFDMS), shown in figure 1(c), contains ferrocene as part of the polymer backbone. In order to separate the influence of the methyl groups in PFDMS, we additionally include poly(ferrocenylmethylsilane) (PFMS) shown in figure 1(d) in our study. Here, one out of the two methyl groups per monomer moiety is substituted by a hydrogen.

For better understanding of the redox-responsiveness of these polymers, we investigated the effect of oxidation on the vibrational properties of ferrocene and ferrocene-containing polymers. We first compare bulk ferrocene to ferrocenium triiodide salt and subsequently study PVFc after oxidation with iodine.

## 2. Experimental

Ferrocene was purchased from Alfa Aesar (purity 99 %) and used without further purification. Ferrocenium triiodide ($FcI_3$) was obtained by stoichiometric mixing of ferrocene and iodine in dichloromethane and subsequent evaporation of the solvent yielding a black powder.

PVFc ($20\,kg\,mol^{-1}$) [31], PFDMS ($16\,kg\,mol^{-1}$ and $88\,kg\,mol^{-1}$) [32] and PFMS ($28\,kg\,mol^{-1}$) [33] were synthesized by anionic polymerization according to procedures described in literature. The molecular weights of the obtained polymers were determined by gel permeation chromatography (GPC). In order to obtain oxidized PVFc, 213 mg (0.84 mmol) iodine were added to a solution of 177 mg PVFc (0.84 mmol corresponding to ferrocene units) in 10 ml THF. After 1 h, 80 ml diethyl ether were added to precipitate the oxidized polymer. 250 mg oxidized PVFc is obtained after washing with diethyl ether and drying in vacuo. As the distribution and character of the iodine counterions (e.g., $I^-$ or $I_3^-$) was not investigated further, we refer to the oxidized compound as $PVFcI_x$.

Vibrational spectra were recorded on the hot neutron spectrometer IN1-Lagrange at the

Institut Laue-Langevin, Grenoble (France). For the neutron scattering experiments the powder samples of $FcI_3$ and $PVFcI_x$ were packed in aluminum foil and rolled into a hollow cylinder of $\approx$ 15 mm diameter. The remaining samples (Fc, PVFc, PFDMS16, PFDMS88, PFMS) were loaded in $3 \times 4$ cm flat aluminum containers of 0.3–0.5 mm thickness. All experiments were performed at a sample temperature of $T = 10$ K controlled by a standard closed-cycle He cryostat. The incident neutron energy $E_i$ was varied by scanning the Bragg angle of different monochromator reflections: Si(111) (5–22 meV), Si(311) (17–70 meV) and Cu(220) (40–500 meV). The downward scattered neutrons from the sample are selected at fixed $E_f = 4.5$ meV and focused on a single detector by a large area PG(002) analyzer covering scattering angles $2\theta$ in the range $\approx$ 21–159°. A typical energy resolution was $\approx$ 2–4 % in energy transfer $\hbar\omega = E_i - E_f$ depending on the selected monochromator reflection and $E_i$. Scattering from the empty cell was subtracted, and the recorded intensity was normalized and corrected for energy dependent monitor efficiency. Subsequently, correction factors from a water sample measurement were applied to compensate for a minor dependency of the monitor flux on the monochromator take-off angle. The resulting intensity is proportional to the dynamic structure factor $S(Q, \omega)$ of the sample, where the magnitude of the scattering vector $Q$ depends on $\omega$ and the scattering angle $2\theta$ which is sampled over the whole $2\theta$-range covered by the analyzer. Finally, from the relationship [34]

$$G(\omega) \propto \frac{\omega}{Q^2} S(Q, \omega) \left(1 - e^{-\hbar\omega/k_B T}\right) \tag{1}$$

the generalized density of states $G(\omega)$ is calculated and will be discussed further in this paper.

## 3. Results & Discussion

In the following, we will present the vibrational spectra collected in this study. We start with a comparison of bulk Fc in its triclinic crystalline phase and various ferrocene-containing polymers. Subsequently, we turn to the effect of oxidation and discuss the oxidized compounds $FcI_3$ and $PVFcI_x$.

### *3.1. Ferrocene and Ferrocene-containing polymers*

The vibrational spectra of triclinic Fc and different ferrocene-containing polymers is shown in figure 2. Vertical lines correspond to maxima in the Fc spectrum for easier comparison. The spectrum for Fc (shown in the bottom part of figure 2) exhibits collective lattice modes in the low energy range up to 100 cm$^{-1}$ as well as local molecular excitations at higher frequencies. It compares well to previous INS measurements published by Kemner *et al.*, where a detailed discussion, assignment and visualization of the individual modes is given [29]. The sketch below the graph in figure 2 indicates the displacements associated with the modes between 180 cm$^{-1}$ and 595 cm$^{-1}$. The broadened, intense peak structure at higher frequencies between 800 and 1100 cm$^{-1}$ is caused by several modes involving internal distortions of the cyclopentadienyl (Cp) rings and has been analyzed in more detail by Kemner *et al.* [29].

Compared to crystalline ferrocene, the spectrum of amorphous PVFc in the middle part of figure 2 shows less detail as expected. However, for this polymer, we expect the vibrational spectrum to bear the most resemblance to bulk Fc as the ferrocene moieties are attached laterally as side chain by a single C—C bond. Indeed the modes at 390, 484, 500 and 595 cm$^{-1}$ are clearly distinguishable in the spectrum of PVFc, and only the mode at 500 cm$^{-1}$ shows a slight shift to higher frequencies and the mode 595 cm$^{-1}$ a slight shift to lower frequencies with respect to bulk Fc. The mode at 180 cm$^{-1}$ corresponding to ring-metal-ring bending with antisymmetric ring tilt is also slightly shifted to lower frequencies and broadened in PVFc. The superposition of multiple modes mostly involving ring distortion and C—H bending leading to the 4 broad peaks between 800 and 1100 cm$^{-1}$ appear even more smeared in PVFc, but not shifted.

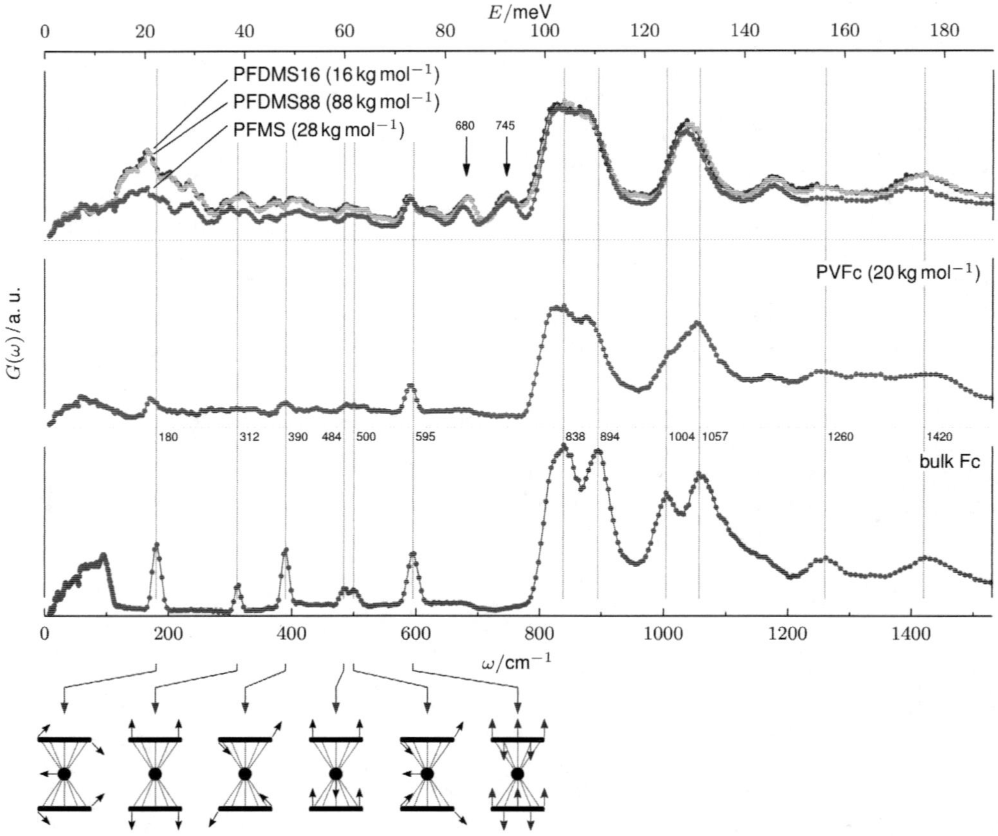


**Figure 2.** Experimental generalized density of states $G(\omega)$ for triclinic ferrocene (Fc), poly(vinylferrocene) (PVFc), poly(ferrocenyldimethylsilane) (PFDMS) with two molecular weights, and poly(ferrocenylmethylsilane) (PFMS). All spectra were recorded at $T = 10\,\mathrm{K}$. Vertical lines show characteristic maxima of the Fc spectrum for easier comparison ($\omega$ given in $\mathrm{cm}^{-1}$). The sketch below the graph indicates the type of distortion associated with some fingerprint modes of the ferrocene molecule.

Interestingly, the mode at $312\,\mathrm{cm}^{-1}$ is not easily visible in the spectrum of PVFc. This mode corresponds to ring-metal-ring stretching involving the collective oscillation of both Cp rings towards the metal, which is here found to be significantly influenced when the Fc moiety is attached to the polymer chain in PVFc. One could speculate that this mode is, due to its large collective displacement, particularly sensitive to the local environment and extremely broadened in the amorphous polymer.

We now come to the polymers PFDMS and PFMS where the ferrocene complex is part of the polymer backbone itself (see figure 1). Their vibrational spectra are shown in top part of figure 2 and are scaled in intensity to overlap in the low frequency range $\omega < 100\,\mathrm{cm}^{-1}$ and around the broad peak at 800–900 $\mathrm{cm}^{-1}$. For PFDMS, two molecular weights of 16 and 88 $\mathrm{kg\,mol}^{-1}$ were investigated. Assuming that the latter broad peak is governed by internal modes of the ferrocene complex, this corresponds to a normalization to the number of monomer units. In this representation, it is obvious that the spectra of PFDMS16 and PFDMS88 are identical and independent of the molecular weight. The spectrum of PFMS differs mostly in the

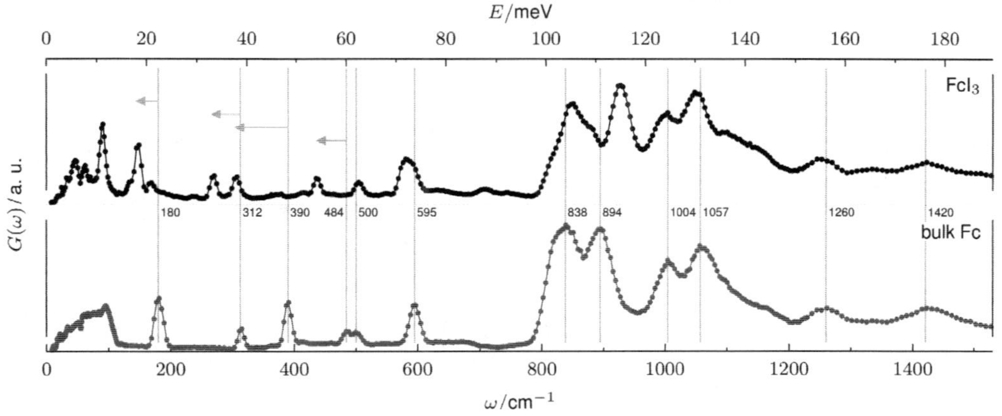


**Figure 3.** Experimental generalized density of states $G(\omega)$ for triclinic ferrocene (Fc), and ferrocenium triiodide ($FcI_3$). All spectra were recorded at $T = 10\,K$. Vertical lines show characteristic maxima of the Fc spectrum for easier comparison ($\omega$ given in $cm^{-1}$).

range 100–300 $cm^{-1}$ where the broad peak structure is less intense, but of similar shape.

As the only difference in the chemical structure of PFMS and PFDMS is the number of methyl groups per monomer, this supports the conclusion that the torsional mode of the methyl groups contributes in this frequency region. A comparison with poly(dimethylsiloxane) (PDMS) [35, 36] shows that the mode around 140 $cm^{-1}$ might be ascribed to the methyl group torsional mode, followed by the 180 $cm^{-1}$ ring-metal-ring bending mode of ferrocene which is shifted to lower frequency.

Moreover, the spectrum of PDMS as discussed by Jayes *et al.* [35] helps to assign the newly appearing peaks at 680 and 745 $cm^{-1}$ (arrows in the top part of figure 2) in PFDMS and PFMS to $CH_3$ rocking modes. However, we may expect a decreased intensity of these modes in PFMS due to the reduced number of methyl groups which is not the case for the right peak at 745 $cm^{-1}$.

Concerning the characteristic modes of the ferrocene complex, we find for PFDMS and PFMS instead of a single ring-metal-ring stretching mode at 312 $cm^{-1}$ a symmetrically split double peak – the same holds for symmetric ring tilt at 390 $cm^{-1}$. We can imagine that the breaking of the rotational symmetry of the Fc complex caused by inclusion in the polymer backbone leads to a splitting of the vibrational modes.

At higher frequencies, the two peaks at 1004 and 1057 $cm^{-1}$ in bulk Fc merge to a single peak in PFDMS and PFMS. They correspond to C—H bending modes in the Cp ring plane which are expected to be strongly modified when the ferrocene molecule is bound as part of a polymer chain.

### *3.2. Oxidized compounds*

In addition to ferrocene and ferrocene-containing polymers, we studied the oxidized compounds $FcI_3$ and $PVFcI_x$ in order to investigate the influence of oxidation on the ferrocene complex. Figure 3 shows the comparison of bulk Fc with $FcI_3$. Both substances are in a crystalline phase. The low energy part of the spectrum $< 100\,cm^{-1}$ shows several sharp peaks for $FcI_3$. One can speculate that this peak structure is caused by optical phonon branches corresponding to low energy out-of-phase vibrations of the $Fc^+$ and $I_3^-$ molecules. The remaining modes between 180 and 484 $cm^{-1}$ differ significantly in $FcI_3$ with respect to Fc. It is consistent with the expectation that the weakening of the ferrocenium ion by a missing electron leads to a shift

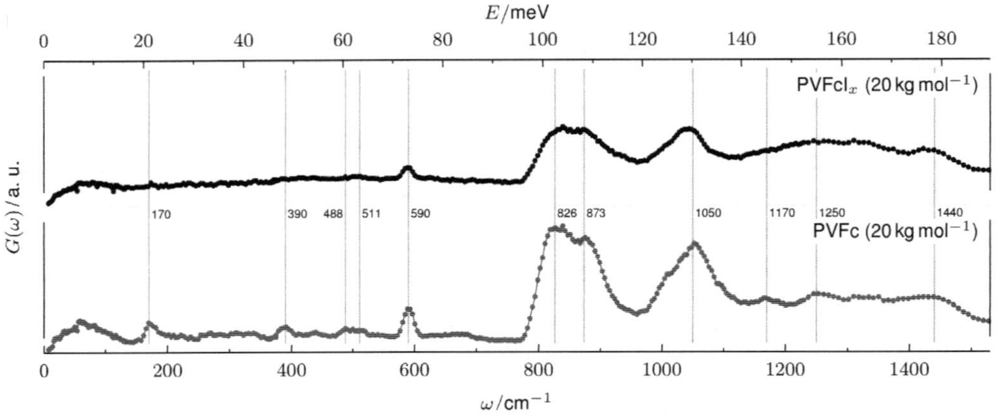


**Figure 4.** Experimental generalized density of states $G(\omega)$ for poly(vinylferrocene) (PVFc), and poly(vinylferrocene) oxidized with iodine (PVFcI$_x$). All spectra were recorded at $T = 10\,\text{K}$. Vertical lines show characteristic maxima of the PVFc spectrum for easier comparison ($\omega$ given in cm$^{-1}$).

to lower frequencies for internal modes involving the $\eta^5$-complex bond. The grey arrows in the figure give a possible assignment of the modes undergoing a shift. Modes at higher frequencies involving deformation of the Cp rings show less differences, with the exception of the peak at $894\,\text{cm}^{-1}$. A more detailed study and phonon calculations of FcI$_3$ will certainly help in giving the correct assignment and interpretation of the influence of oxidation on the vibrational spectrum of ferrocene. This work is currently in progress.

In contrast to Fc and FcI$_3$, the comparison of the ferrocene-containing polymer PVFc and its iodine oxidized compound PVFcI$_x$ shown in figure 4 exhibits significantly less detail. For the oxidized polymer, the background is much higher and aside from the broad ring deformation modes above $800\,\text{cm}^{-1}$ which are left mostly uninfluenced by the oxidation, only the out of plane ring deformation mode at $590\,\text{cm}^{-1}$ is distinguishable in the spectrum of PVFcI$_x$. This mode is, as in the case of FcI$_3$, not shifted with respect to the unoxidized compound. The peak at $170\,\text{cm}^{-1}$ corresponding to ring-metal-ring bending with antisymmetric ring tilt is not distinguishable in the spectrum of PVFcI$_x$ anymore – the comparison of Fc and FcI$_3$ has already shown that this mode is significantly shifted to lower frequencies and is thus probably not separable from the low energy excitations.

## 4. Conclusion

The use of inelastic neutron spectroscopy enabled us to investigate the vibrational spectrum of the ferrocene complex in different redox-responsive ferrocene-containing polymers. When the ferrocene moiety is attached as a side group as in PVFc, we found significant broadening of the ring-metal-ring bending with antisymmetric ring tilt and no sign of the ring-metal-ring stretching mode. With these exceptions, no shift of the vibrational fingerprint modes in the range 100–800 cm$^{-1}$ was detected.

In contrast to this, a more dramatic influence on the vibrational modes was observed when the ferrocene is part of the polymer backbone as in PFDMS. The interpretation of these data is complicated by significant contributions from methyl groups of the polymer. However, the study of PFMS and comparison to previous results for PDMS helped in identifying methyl group torsion contributions.

In case of oxidation of the ferrocene complex, the weakening of the coordination bond in $FcI_3$ leads to a significant shift of multiple fingerprint modes to lower frequencies. A DFT study to substantiate this observation and give an unambiguous assignment of the modes is currently in progress.

**Acknowledgments**
The Institut Laue-Langevin (Grenoble, France) is acknowledged for the allocation of beamtime on IN1-Lagrange. The authors would like to thank Andreas Weber and Monica Jimenez-Ruiz for their assistance during the experiment. J. E., M. G. and B. S. thank the Landes-Offensive zur Entwicklung Wissenschaftlich-ökonomischer Exzellenz (LOEWE Soft Control) for ongoing financial support of this work.

**References**
[1] Stuart M A C, Huck W T S, Genzer J, Müller M, Ober C, Stamm M, Sukhorukov G B, Szleifer I, Tsukruk V V, Urban M, Winnik F, Zauscher S, Luzinov I and Minko S 2010 *Nat. Mater.* **9** 101–13
[2] Chen T, Ferris R, Zhang J, Ducker R and Zauscher S 2010 *Prog. Polym. Sci.* **35** 94–112
[3] Hu J, Meng H, Li G and Ibekwe S I 2012 *Smart Mater. Struct.* **21** 053001
[4] Lee H, Pietrasik J, Sheiko S S and Matyjaszewski K 2010 *Prog. Polym. Sci.* **35** 24–44
[5] Minko S 2006 *Polym. Rev.* **46** 397–420
[6] Lallana E and Tirelli N 2013 *Macromol. Chem. Phys.* **214** 143–58
[7] Ramanathan M, Tseng Y C, Ariga K and Darling S B 2013 *J. Mater. Chem. C* **1** 2080–91
[8] Abd-El-Aziz A S, Agatemor C and Etkin N 2014 *Macromol. Rapid Commun.* **35** 513–559
[9] Whittell G R, Hager M D, Schubert U S and Manners I 2011 *Nat. Mater.* **10** 176–88
[10] Bellas V and Rehahn M 2007 *Angew. Chem., Int. Ed.* **46** 5082–104
[11] Whittell G and Manners I 2007 *Adv. Mater.* **19** 3439–68
[12] Manners I 2004 *Synthetic Metal-Containing Polymers* (Weinheim: Wiley)
[13] Hempenius M A, Cirmi C, Lo Savio F, Song J and Vancso G J 2010 *Macromol. Rapid Commun.* **31** 772–83
[14] Nakahata M, Takashima Y, Yamaguchi H and Harada A 2011 *Nature Commun.* **2** 511
[15] Staff R H, Gallei M, Mazurowski M, Rehahn M, Berger R, Landfester K and Crespy D 2012 *ACS Nano* **6** 9042–9
[16] Elbert J, Gallei M, Rüttiger C, Brunsen A, Didzoleit H, Stühn B and Rehahn M 2013 *Organometallics* **32** 5873–8
[17] Elbert J, Mersini J, Vilbrandt N, Lederle C, Kraska M, Gallei M, Stühn B, Plenio H and Rehahn M 2013 *Macromolecules* **46** 4255–67
[18] Elbert J, Krohm F, Rüttiger C, Kienle S, Didzoleit H, Balzer B N, Hugel T, Stühn B, Gallei M and Brunsen A 2013 *Adv. Funct. Mater.* **24** 1591–1601
[19] Rider D A and Manners I 2007 *Polym. Rev.* **47** 165–95
[20] Ahmed R, Priimagi A, Faul C F J and Manners I 2012 *Adv. Mater.* **24** 926–31
[21] Schacher F H, Elbert J, Patra S K, Yusoff S F M, Winnik M A and Manners I 2012 *Chem.-Eur. J.* **18** 517–25
[22] Schacher F H, Bellas V, Winnik M A and Manners I 2013 *Soft Matter* **9** 8569–78
[23] Zhang M, Rupar P A, Feng C, Lin K, Lunn D J, Oliver A, Nunns A, Whittell G R, Manners I and Winnik M A 2013 *Macromolecules* **46** 1296–304

[24] Tonhauser C, Mazurowski M, Rehahn M, Gallei M and Frey H 2012 *Macromolecules* **45** 3409–18

[25] Morsbach J, Natalello A, Elbert J, Winzen S, Kroeger A, Frey H and Gallei M 2013 *Organometallics* **32** 6033–9

[26] Kraska M, Gallei M, Stühn B and Rehahn M 2013 *Langmuir* **29** 8284–91

[27] Gallei M, Schmidt B V K J, Klein R and Rehahn M 2009 *Macromol. Rapid Commun.* **30** 1463–9

[28] Schmidt B V K J, Elbert J, Barner-Kowollik C and Gallei M 2014 *Macromol. Rapid Commun.* **35** 708–714

[29] Kemner E, de Schepper M, Kearley G and Jayasooriya U 2000 *J. Chem. Phys.* **112** 10926–29

[30] Kemner E, Overweg A R, Jayasooriya U A, Parker S F, de Schepper I M and Kearley G J 2002 *Appl. Phys. A: Mater. Sci. Process.* **74** S1368–S1370

[31] Gallei M, R Klein R and Rehahn M 2010 *Macromolecules* **43** 1844–54

[32] Kloninger C and Rehahn M 2004 *Macromolecules* **37** 1720–7

[33] Du V A and Manners I 2013 *Macromolecules* **46** 4742–53

[34] Lovesey S 1984 *Theory of Neutron Scattering from Condensed Matter* (*International Series of Monographs on Physics* vol 1) (Oxford: Clarendon Press)

[35] Jayes L, Hard A P, Séné C, Parker S F and Jayasooriya U A 2003 *Anal. Chem.* **75** 742–6

[36] Hard A P, Parker S F and Jayasooriya U A 2007 *Appl. Spectrosc.* **61** 314–20

# Hydrogen on graphene investigated by inelastic neutron scattering

**C Cavallari [1,2], D Pontiroli[2], M Jiménez-Ruiz[1], A Ivanov [1], M Mazzani [2], M Gaboardi [2], M Aramini [2], M Brunelli[1], M Riccò [2] and S Rols [1,3]**

[1] Institut Laue Langevin, 6 rue J. Horowitz, 38042 Grenoble, France
[2] Dipartimento di Fisica e Scienze della Terra, Università degli Studi di Parma Parco Area delle Scienze 7/A, 43124 Parma, Italia

Email: rols@ill.fr, cavallari@ill.fr

**Abstract**. Inelastic neutron scattering experiments and *ab-initio* calculations have been used to investigate the vibrational modes, in a wide energy region between 0 and 200 meV, of hydrogenated graphene produced by chemical method. The results show the presence of atomic hydrogen chemisorbed at the graphene surface. At 10 K, the measured high energy density of states is remarkably similar to that of hydrogenated ball-milled graphite, in which hydrogen is most likely bonded to C atoms at the edges. In fact, in both hydrogenated graphene and hydrogenated ball-milled graphite, the high frequency modes (100-200 meV) show strong similarities with the C-H bending modes of the coronene molecule, in which hydrogen is bonded at the edges. This hypothesis has been supported by *ab-initio* calculations.

## 1. Introduction

The hydrogen interaction with graphene is attracting considerable attention from both fundamental and applicative implications. The physico-chemical understanding of the peculiar state of hydrogen at the graphene surface is highly relevant in many different fields, ranging from interstellar chemistry to the possibility of tailoring its electric and magnetic properties, just to mention few examples.

Most of the recent efforts regarding H adsorption have been stimulated by the prospect of using graphene-based compounds as hydrogen storage media. In the recent years, the massive production of graphene exploiting the thermal exfoliation of graphite oxide has opened the route towards new practical applications of graphene e.g in the field of hydrogen storage. From the fundamental point of view, the gram scale graphene synthesis allowed experimental investigation using bulk techniques that usually require large amount of samples [1].

So far, the interaction between hydrogen and graphene has been studied extensively especially from the theoretical point of view.

[3] To whom any correspondence should be addressed

Journal of Physics: Conference Series **554** (2014) 012009 doi:10.1088/1742-6596/554/1/012009

From the experimental point of view, atomic hydrogen was found to passivate some fractions of dangling bonds [2]. However, a variety of C-H chemical bonds occur in different carbon materials and a general tendency for H to saturate π-bonds on the plane, converting $sp^2$ to $sp^3$ carbon was also observed [3].

The precise mechanism of H adsorption is still unknown and according to experiments, the H state at the graphene surface, as well as the stability of H aggregates, strongly depend on the chemical state of the surface and it is therefore extremely sensitive to the graphene synthesis, manipulation and history.

In this paper we present a study of the hydrogen dynamics at the graphene surface, with inelastic neutron scattering (INS) and DFT molecular dynamics (MD) simulations. Graphene samples were synthesized by thermal exfoliation of graphite oxide (TEGO) and then exposed to hydrogen flux at 1073 K. Graphene (TEGO) and nanographite (ball-milled graphite) have been investigated in both the pristine conditions and after hydrogenation, in order to evidence, in particular, the peculiarities of hydrogen at the graphene surface.

Sample synthesis and preparation, neutron experiments and computation methods are described in Section 2. The discussion of the results and the conclusion are given in Section 3 and Section 4, respectively.

## 2. Experimental procedure

### *2.1. Synthesis and preparation*

Graphene samples under investigation were produced by thermal exfoliation of graphite oxide [4].

Commercial highly pure graphite powder (RW-A grade, purchased from SGL Carbon group) was oxidized following the Brodie method, to obtain graphite oxide (GO). The reduction of GO is reached by thermal exfoliation at 1320 K, under vacuum for 30 min. TEGO samples have been previously characterized with Transmission Electron Microscopy (TEM), Scanning Electron Microscopy (SEM) and Selected-Area Electron Diffraction (SAED). The complete characterization has been reported in [3] and [1] and clearly demonstrated that the samples contain both single and few layers structures of $sp^2$ carbon [5]. Nanographite was obtained by mechanical ball-milling of graphite in strict oxygen and moisture free conditions. Highly pure commercial graphite was measured as a reference.

Hydrogenated samples were obtained by thermal treatment of pristine TEGO in a hydrogen flux of 110 ml/min, keeping the maximum temperature of 1073 K for one hour. Hydrogenated TEGO (H-Graphene) was the principal subject of our investigations.

All the synthesis steps, preparation and hydrogenation process, as well as the subsequent sample handling and storage were done in argon atmosphere (oxygen and moisture levels below 1 ppm), in order to prevent any possible contamination of the samples with air.

### *2.2. Inelastic Neutron Scattering*

In order to probe the local hydrogen environment, we have performed inelastic neutron scattering (INS) experiments. Vibrational spectroscopy based on INS is a very powerful method to provide information about the dynamical properties of matter and offers several advantages, in particular the very high sensitivity to modes involving hydrogen motion and the absence of symmetry selection rules. Hydrogen has a very strong incoherent cross section, which is more than one order of magnitude larger than the total scattering cross section of carbon. In the case of inelastic scattering for hydrogenated compounds, all scattering can be assumed incoherent and the double differential cross section is then directly related to the vibrational density of states and can be written as [6]:

$$\left.\frac{\partial^2\sigma}{d\Omega dE}\right|_{inc} = \frac{N}{8\pi}\frac{k_f}{k_i}\langle n(\omega)+\frac{1}{2}\pm\frac{1}{2}\rangle\frac{Q^2}{\omega}\sum_{j=1}^{3N-6}\delta(\omega-\omega_j(\vec{q}))\times\sum_{i=1}^{N}\frac{\sigma_i}{M_i}|\hat{Q}\cdot\widehat{e_{ij}}|^2\, e^{-2W(\vec{Q})_i}$$

where $k_i$ and $k_f$ are the incident and scattered neutron wave vector, $\vec{Q}$ is the momentum transfer, $n(\omega)$ is the temperature dependent Bose population factor, $e^{-2W(\vec{Q})}$ is the Debye-Waller factor.

Journal of Physics: Conference Series **554** (2014) 012009 doi:10.1088/1742-6596/554/1/012009

Inelastic Neutron Scattering experiments were performed using two complementary techniques, Time-of-Flight spectroscopy and filter-analyzer neutron spectroscopy. The combined use of these techniques allowed us to measure the vibrational density of states in a very wide energy region, ranging from 0 meV up to hundreds of meV. We used the hot neutron spectrometer IN1-BeF, at the Institut Laue Langevin (Grenoble, France), with the Beryllium filter option: a block of polycrystalline Be, cooled to 20 K, scatters out all neutrons having final energy higher than 3.5 meV. A Cu(220) monochromator was used, limiting the neutron energy in the low part of the spectrum at 35 meV and providing an energy resolution of approximately 4% in the energy range probed (up to 200 meV). The samples were held at 10 K using a standard Orange cryostat.

In the extent that $k_f$ is much smaller than $k_i$, that is the case of the Beryllium filter analyzer spectrometer IN1BeF, the observed intensity in a low-temperature experiment is directly proportional to the phonon density of states.

The inelastic neutron scattering at lower energies was performed using the Fermi chopper thermal neutron time-of-flight spectrometer IN4C, also at the Institut Laue Langevin.

From time-of-flight data, the Generalized vibrational Density of States (GDOS) can be derived in an extensive (0-200 meV) energy scale using a relatively large neutron incident wavelength (in this work we used $\lambda$=2.4 Å) in up-scattering mode (neutron energy gain) and relatively high temperature (320 K). GDOS were obtained from time-of-flight data after proper treatment, corrections and normalization to the mass of the samples, as described in the ref [7].

In both experiments the samples (about 0.5 g) were sealed inside standard Al flat container with an Indium O-ring. Cell-filling and sample-handling for the neutron experiments were carried out inside a glove-box under pure argon atmosphere.

The simulations of poly-aromatic hydrocarbons with chemical formula $C_{6n^2}H_{6n}$ with n=2, 3 and 4 were performed by taking a one-layer model of (0001) graphite basal plane with H covalently bonded at the external edges. Ab-initio molecular dynamics trajectories were produced using the frozen-core all-electron plane wave (PAW) method, as implemented in the Vienna *ab initio* DFT code VASP [8] with the GGA-PBE approximation [9]. Fully relaxed structures were used as input for the trajectory productions in the MD runs. A 2 ps thermal equilibration step was performed initially at 300 K in the NVT ensemble (time step 1.0 fs) using the Berendsen thermostat with a velocity scaling each 2 steps. The equilibration runs were followed by a production run of 10 ps performed also in the NVT ensemble and using the Nose thermostat, with a coupling constant of 96 THz. We used the nMOLDYN program [10] to compute the GDOS as the time-Fourier transform of the velocity autocorrelation function (VACF), from the coordinates of the atoms at each step of the *ab-initio* molecular dynamics (MD) trajectories. The computation of the VACF using an incoherent weight allows extraction of the GDOS of the H atoms directly. Finally, the simulated GDOS were convoluted with a Gaussian function, to take into account the instrumental resolution.

## 3. Discussion and Results

As preliminary characterization, the GDOS of graphite, nanographite and graphene are reported in Figure 1a. The GDOS of graphite shows the well-known distinct and well-defined features corresponding to flat regions of the dispersion curves [11]. The reduced dimensions of the grains in nanographite and graphene have strong effects on the dynamical properties: peaks in the density of states progressively broaden, with an increased proportion in the low frequency part of the spectrum. This reveals the decrease in dimensionality of the system by introduction of turbostratic disorder, up to a complete loss of out of plane coherence in defective graphene.

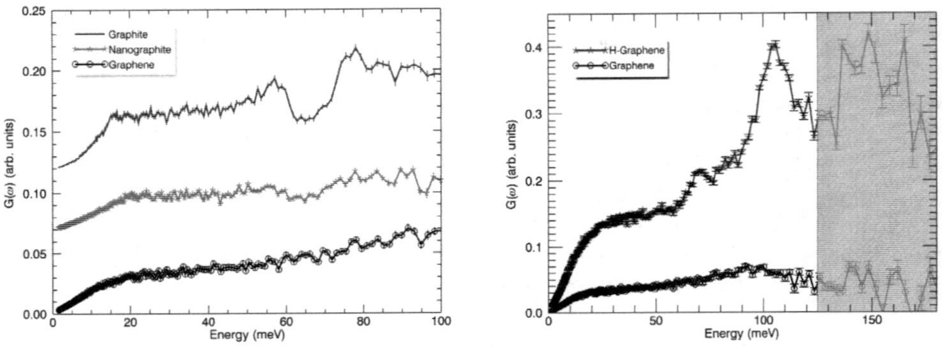


**Figure 1a (left) Generalized Density of States (GDOS) measured at 320K and 2.4Å of graphite (solid blue line, in the upper part of the figure), ball-milled graphite (nanographite, in red stars, in the middle) and as-prepared TEGO (black circle, in the bottom part of the figure). Figure 1b (right) Generalized Density of States (GDOS) measured in the same conditions of as-prepared TEGO (black circles) and H-TEGO (blue stars). In H-TEGO features at high frequencies are mainly due to internal vibration of H. The low frequency region shows a featureless intensity, assigned to lattice vibration of reduced coherence extension, originating from phonons in defective graphene. Data in the high frequency part of the spectrum have been overshadowed because the low thermal population of these modes at 320 K does not allow to rigorously discuss to features observed in up-scattering mode.**

Figure 1b shows the effect of the hydrogenation. The enhancement in intensity observed in the GDOS after hydrogenation of TEGO denotes an increased quantity of hydrogen. It is possible to distinguish two different frequency domains, distributed over a broad energy region. In the low frequency region of the spectra ([0, 50 meV]) no sharp features are observed. A linear increase of the GDOS is observed up to 20 meV, and upon this value it becomes flat. This range mostly features out–of–plane modes of the graphene surface. The constant shape but increased intensity of the GDOS in this range from as prepared TEGO to H–TEGO reflects that the hydrogen atoms follow adiabatically the plane motion for these modes, so-called riding modes.

By contrast, additional bands, located around 70, 100 and 150 meV are observed in the high frequency region of H-TEGO. These features are completely compatible with the C-H bending modes in polycyclic aromatic hydrocarbons, as in coronene or corannulene [12,13], witnessing the effectiveness of defects in dissociating the molecule of hydrogen and trapping atomic hydrogen in covalent bonds. The very low thermal population of the high energy modes at the temperature we performed the experiment (320 K) in up-scattering mode does not allow to rigorously discuss to features observed, in particular the band located around 150 meV.

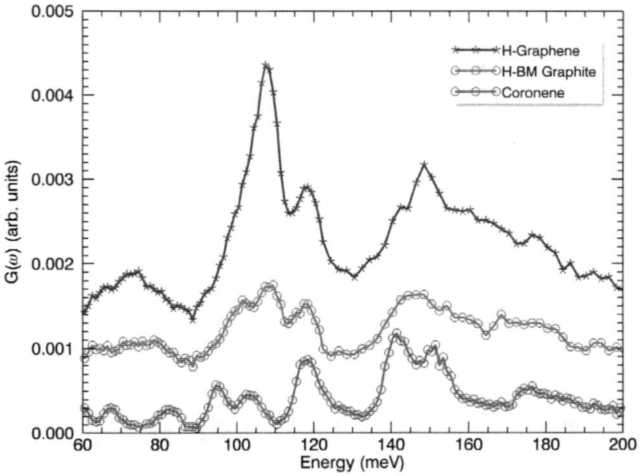


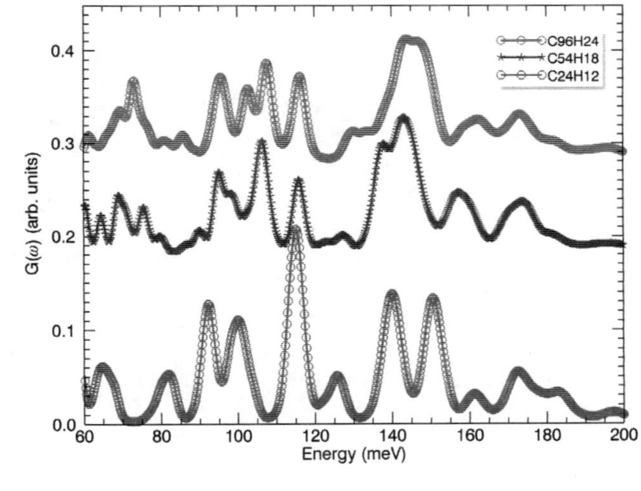


**Figure 2 (top): Inelastic scattering spectra of H-graphene (blue stars) and H-BM graphite (hydrogenated nanographite in red circles) measured at 10K using the high resolution Be-filter hot neutron spectrometer IN1.**

**Figure 3 (bottom): Simulated INS spectra of coronene and larger hydrogenated aromatic molecules of the type $C_{6n^2}H_{6n}$ (in this work, n=2, 3, 4 have been considered. The case n=2 corresponds to coronene).**

To get a deeper insight into the nature of the C-H vibrational bands of hydrogenated graphene, high resolution spectra of the same sample, together with hydrogenated nanographite and coronene were obtained using the IN1Be filter spectrometer and are shown in Figure 2. The internal complex structures of the H-graphene vibrational bands already observed in the TOF spectra, especially those of the peak located around 110 meV, are now better resolved.

It is possible to highlight important similarities between the spectra of H-nanographite and H-graphene, especially concerning the position in frequency of the peaks, which can be attributed to common dynamical features. In particular, the spectra can be decomposed into three zones: the first below 90 meV features a broad peak at ~ 70 meV, the second between 90 and 130 meV features a complex structure formed of several peaks with a dominant contribution at ~107 meV and a secondary maximum at ~117 meV, the third above 130 meV is composed of a broad asymmetric band with maxima located at ~ 145 meV. Similar features are found in the coronene data, although the molecular nature of the latter being responsible for the clearer definition of the bands. In particular, the peak at ~105 meV is well reproduced.

Calculations for coronene $C_{24}H_{12}$ and two larger hydrogenated graphene fragments $C_{6n^2}H_{6n}$ with n=2 and 3 are presented in Figure 3. A very good agreement between the simulated and experimental GDOS is obtained for coronene. When increasing the size of the hydrogenated aromatic molecules from $C_{24}H_{12}$ to $C_{96}H_{24}$, one observes a progressive evolution of the coronene molecular spectrum towards a more complex spectrum, with lattice-like character: the well defined peaks merge into frequency zones, featuring complex structures and gaps. The three zones description of the data can be applied to the $C_{96}H_{24}$ simulated GDOS, and the overall shape of the latter can account for most of the features observed in the H-nanographite spectrum. This observation suggests that H atoms are located at the edges of the nanodomain regions of the H-nanographite sample, and that single H saturation per edge C atoms applies.

By contrast, our simulations fail to reproduce the main feature of the H-TEGO GDOS, e.g. no dominant peak at 107 meV is calculated. This means that a model in which H atoms are located only at the external borders of the graphene flake does not describe the bending mode region of the H-graphene spectrum in a complete way. Compared to nanographite, thermal exfoliated graphite oxide is well known to have a much more extended surface area and to be characterized by in-plane defects, produced during the synthesis process. The extra dynamical features in H-graphene, which appear as over-intensity located in particular at the 107 meV band, can be attributed to H covalently bonded to in-plane defects.

## 4. Conclusions

In this paper, we have presented the results of the first investigation of the local hydrogen environment in hydrogenated graphene, produced by chemical methods, using neutron spectroscopy and DFT simulations.

Our work evidences the effectiveness of defects in dissociating the hydrogen molecules and trapping atomic hydrogen with covalent bonds.

We found that the frequencies of the C-H bending modes in the neutron spectra are strongly comparable to the ones of polycyclic aromatic molecules as coronene, in which H is located at the external edges. We isolated a contribution at 107 meV that we associate with H atoms bonded at the surface of the graphene plane.

INS, in combination with DFT calculations, constitutes a unique way to efficiently probe the H states at the graphene surface and has allowed for unraveling its specific configurations at a local scale even in a very complex carbon matrix.

Authors thank ILL for neutron beam-time allocations and ILL C-Lab for simulation support.

## References

[1] Gaboardi M, Bliersbach A, Bertoni G, Aramini M, Vlahopoulou G, Pontiroli D, Mauron Ph, Magnani G, Salviati G, Zuttel A and Riccò M 2014 *J. Chem. Mat. A* **2** 1039

[2] Horneaker L, Rauls E, Xu W, Sljivancanin Z, Otero R, Stensgaard I, Lægsgaard E, Hammer B and Besenbacher F 2006 *Phys. Rev. Lett* **97** 186102

[3] Riccò M., Pontiroli D., Mazzani M., Choucair M., Stride J.A. and Yazyev O.V. 2011 *NanoLett.* **11 (11)**,4919-4922

[4] Park S and Ruoff R S 2009 *Nat. Nanotechnol.* **4** 217

[5] Hernandez Y, Nicolosi V, Lotya M, Blighe F M, Sun Z, De S, McGovern I T, Holland B, Byrne M, Gun'Ko Y K, Boland J J, Niraj P, Duesberg G, Krishnamurthy S, Goodhue R, Hutchison J, Scardaci V, Ferrari A C and Coleman J N 2008 *Nat. Nanotechnol.* **3** 563

[6] Lovesey S W 1984 *Theory of Neutron Scattering from Condensed Matter* (Oxford: Clarendon Press)

[7] Bousige C, Rols S, Cambedouzou J, Verberck B, Pekker S, Kovats E, Durkò G, Jalsovsky

Journal of Physics: Conference Series **554** (2014) 012009 doi:10.1088/1742-6596/554/1/012009

I, Pellegrini E and Launois P 2010 *Phys. Rev. B* **82** 195413
[8] Kresse G and Furthmüller J 1996 *Phys. Rev. B* **54** 11169
[9] Perdew J P, Burke K and Ernzerhof M 1996 *Phys. Rev. Lett.* **77** 3865
[10] Hinsen K, Pellegrini E, Stachura S and Kneller G R 2012 *J. Comput. Chem.* **33** 2043
[11] Nicklow R, Wakabayashi N and Smith H G 1972 *Phys. Rev. B* **5** 4951
[12] Mitchell P C H, Ramirez-Cuesta A J, Parker S F, Tomkinson J and Thompsett D 2003 *J Phys Chem B* **107** 6838
[13] Le Parc R, Hermet P, Rols. S, Maurin D, Alvarez L, Ivanov A, Quimby J.M, Hanley C G, Scott LT and Bantignies J L 2012 *J Phys Chem C* **116** 25089

 doi:10.1088/1742-6596/554/1/012010

# An inelastic incoherent neutron scattering study of the internal dynamics of ergocalciferol and cholesterol

**K Holderna-Natkaniec[1], I Natkaniec[1,2], Z Fojud[1] and B Peplinska[1]**

[1] Department of Physics Adam Mickiewicz University 61-617 Poznan, Poland
[2] Frank Laboratory of Neutron Physics Joint Institute of Nuclear Research, 141980 Dubna, Russian Federation

E-mail: natkanie@amu.edu.pl

**Abstract.** A model of the methyl group dynamics of two steroid compounds: cholesterol and ergocalciferol was proposed on the grounds of complementary studies performed by inelastic incoherent neutron scattering, nuclear magnetic resonance and quantum chemistry calculation methods.

## 1. Introduction

Steroids are biologically active compounds, derivatives of cholesterol, playing a number of important functions in living organisms. Cholesterol (Chol) is the precursor of bile acids, steroid hormones and provitamin $D_2$ (ergosterol). Ergosterol (ERG) is a component of fungal cell membranes and yeast. It serves the same function as cholesterol in animal cells, which is important in human diet. Irradiation of ERG with ultraviolet light leads to formation of ergocalciferol (ECF) - an isomeric member of vitamin $D_2$. The melting points of ECF, ERG and Chol are 388, 435 and 421 K, respectively. The DSC data of Chol also indicate a phase transition at a temperature close to 310 K.

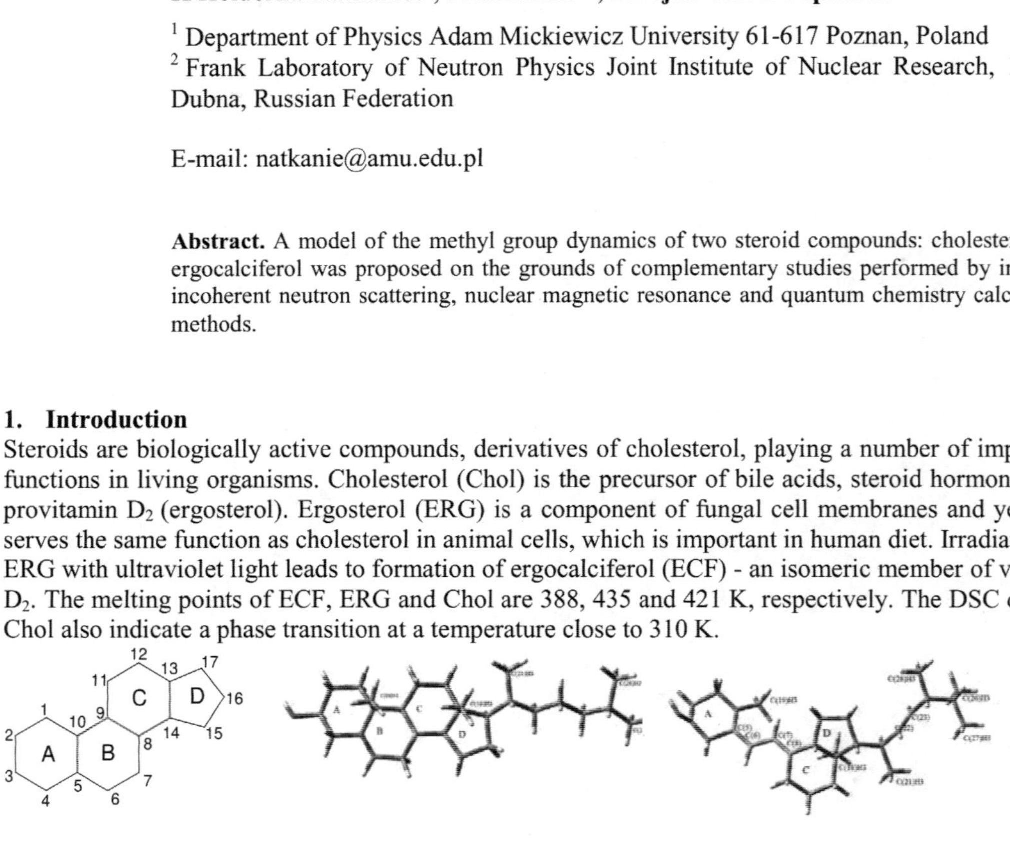


**Figure 1**a. Numbering of the cyclopentano-perhydrophenanthrene atoms b. Structure of cholesterol c. Structure of ergocalciferol

The structure of the Chol (cholest–5–en–3β–ol) molecule of $C_{27}H_{46}O$ stoichiometry (Figure 1b) is known from steroid biochemistry and pharmacology as to consist of a four ring cyclopentanoperhydrophenanthrene moiety, of which three are six-membered cyclohexane rings (denoted A, B, C) and one is a five-membered carbon ring (named D) (Figure 1a)[1]. Rings B and C in the steroid molecules are stable, while rings A and D undergo conformational changes. These changes are directly related to the transportation of steroid hormones by proteins to the cell nucleus. The most important group that is critical for hormonal activity is the 4–en–3–one moiety in ring A. The side chain ($C_5$), attached to C(17) and substituted by three methyl groups (C(21)$H_3$, C(26)$H_3$, C(27)$H_3$) also influences biological activity.

The structure of ERG ($C_{28}H_{44}O \cdot H_2O$) (commonly known as provitamin $D_2$) -differs from that of Chol in having two additional double bonds, one in the steroid four-ring structure at C(7)=C(8) and the other in the alkyl side chain at C(22)=C(23). In comparison with the structure of Chol, ERG also has an additional methyl group attached to the alkyl chain at C(24) [3, 4, 6-10].

The isolated molecule of ECF (vitamin $D_2$) consists of only rings A, C, D of the four-ring structure characteristic of steroid compounds. Ring A is a flattened chair structure, ring C is regular chair, and rings C and D are *trans* fused. The diene system between rings A and C is planar and is formed by the following bonds: C(4)–C(5)=C(6) –C(7)=C(8)–C(9)–C(10). Moreover, the 17β side chain curls up at C(24), and atoms C(20)–C(22)=C(23)–C(24) lie in the plane. The structure of an isolated molecule of ECF is presented in Figure 1c [5, 11].

The internal vibrations of ECF and Chol are the subject of our investigations performed by inelastic incoherent neutron spectroscopy (IINS) and quantum chemistry (QC) modelling. The purpose of the studies was to find out how the substitution of the side-chain substituent by methyl groups affects the vibrational spectra, especially taking into regard the results of our earlier studies of pregnanes and androstanes [12-17]. These results were discussed together with those of internal dynamics of methyl groups from $^1$H NMR studies [18-24]. The performed studies suggest that the sequence of onset of particular methyl groups reorientation were affected by the conformation of the A or D rings, or the torsional angle C(13)–C(17)–C(20)–O. Moreover, the temperature dependence of the spin-lattice relaxation time for Chol obtained from the studies permitted Andrew [24] to indicate the presence of at last two relaxation processes. The weaker one is attributed to the reorientation of one methyl group about the $C_3$ axis, and the stronger relaxation process to the reorientation of the remaining four methyl groups.

The aims of the study are:

1. Estimation of the barrier heights for intramolecular reorientations of the methyl groups by quantum chemistry methods in the isolated molecule approximation in order to propose the sequence of onset of their reorientation,
2. Attempt to explain the distribution of relaxation processes,
3. Determination of the activation energies of reorientations of ECF and Chol methyl groups.

Interpretation of the neutron vibrational spectra of ECF and Chol may be performed versus the results of quantum chemical (QC) calculations and discussed with the data of temperature studies of internal dynamics performed by $^1$H NMR method.

In neutron spectroscopy, according to the scattering law, the intensity of each band is proportional to the cross-section $\sigma$ for inelastic neutron scattering. For hydrogen atoms $\sigma$ is 82.2 barn, which is much greater than the 5.55 and 4.23 barn given for carbon and oxygen [26], respectively. Therefore, the vibrational modes in which hydrogen atoms are involved give high intensity bands. For this reason, neutron spectroscopy is a very attractive method with which to study organic chemical compounds, in particular those built of such groups as $CH_3$, $NH_2$ and OH. From proton magnetic resonance studies one can obtain information on internal mobility of the functional groups.

## 2. Experimental

The inelastic incoherent neutron scattering (IINS) spectra were recorded using the inverted geometry time-of-flight spectrometer NERA-PR operating at the pulsed reactor IBR-2 in Dubna (Russia) [27]. The IINS spectra were obtained after summation of signals from 16 detectors, subtraction of the phonons arising from the cryostat and the thin aluminium cassette ($m_{sample}$ =15 g ), and were then converted in the approximation of one-phonon scattering into the generalised function of the phonon density of states $G_{exp}(\nu)$. They were interpreted using the results of DFT / B3LYP/6-311G**

calculations performed using GAUSSIAN 09 [28]. The QC calculations of the energy of the investigated systems *versus* the methyl group orientation around their three-fold symmetry axis of the C–C bond, were performed by the semi-empirical PM3 method (as it is less time consuming) [29, 30], after optimisation of the available crystallographic data using Gaussian 09 [28].
The DFT calculated IR vibrational frequencies and intensities of the normal modes were compared with the experimental ones, in order to check the quality of the optimised structure. The IR spectra were recorded as KBr discs of 1:100 molar mass, using a Bruker IFS 66/s spectrometer with 1 $cm^{-1}$ resolution in the spectral range from 600 to 4000 $cm^{-1}$, 1024 accumulations, with Blackman-Harris apodisation and base line correction.

The molecular dynamics of the protons in ECF have been analysed on the basis of the temperature measurements of the spin-lattice $T_1$ relaxation times as well as the slope line width and the second moment of the absorption signal. The measurements of $T_1$ were performed by the saturation method with the pulse sequence $n \cdot \pi/2 - \tau - \pi/2$ on a Bruker CXP spectrometer operating at a frequency of 200 MHz. The continuous wave $^1H$ NMR experiment was performed using a home-made spectrometer operating in the double modulation system. The first derivative of the absorption signal was recorded by the sweep of the operating resonance frequency of the autodyne generator in the range (24900, 25050) kHz at a constant magnetic field of 0.587 T, as $^{19}F$ stabilization takes place.

The title steroid compounds were purchased from Aldrich, and used as supplied. The sample under study (1 g) was degassed for several hours in a glass ampoule and sealed under vacuum.

### *2.1. Methodology*

Optimization of the structure is necessary for determination of the vibrational spectrum. It was performed for the molecule with the geometry taken from the crystal structure. The geometry optimization and frequency calculation were performed with the Opt=VeryTight and Integral(Grid=UltraFine) options, respectively. They were calculated by the GAUSSIAN 09 package [28] using the 6–311G** basis set and the B3LYP density functional. The building of the molecule was modelled in vacuum. DFT gives the bond lengths of a large set of molecules with a precision of 1-2%. DFT predicts the vibrational frequencies of a broad range of molecules within 5-10% accuracy. The *ab initio* harmonic vibrational frequencies are typically larger than the fundamentals observed experimentally. The sources of this disagreement are as follows: neglect of anharmonicity, incomplete incorporation of electron correlation and use of a finite basis set. Therefore the scaling factor was determined.

The professionally produced package from *ab initio* quantum chemical calculation provides information on the frequencies of normal modes of the studied system and the amplitudes of the atom displacements from the equilibrium positions. The data permit calculation of the neutron scattering intensity for each mode. It can be realized by the a-Climax [33] programme used for modeling of the neutron scattering function $S(Q, \omega)$ at the TOSCA spectrometer [32]. The intensities in the spectrum of single phonon scattering calculated by a-Climax, are presented in the form of the δ function. Thus, taking into regard the geometry of the TOSCA and NERA spectrometers it is possible to calculate the generalized phonon density of state function $G_{cal}(\nu)$ averaged over all scattering angles at the NERA spectrometer. To enable comparison of the quantum chemistry calculation with the experimental data, the δ function of phonon density of state (frequency and intensity) is convoluted with the resolution power of the spectrometer used with the help of the program RESOL [31] to obtain the $G_{cal}(\nu)$ function.

The energy profile was obtained by QC calculations of the energy of the investigated systems *versus* the methyl group orientation around their three-fold symmetry axis of the C–C bond were performed with a step size of 10° by the semi-empirical PM3 method (as it is less time consuming) [29, 30], after optimisation of the available crystallographic data using Gaussian 09 package [28].
The IR vibrational frequencies and intensities of normal modes calculated by DFT method were compared with the experimental ones, to check the quality of the optimised structure.

## 3. Results and discussion

*3.1 G(ν)*

The phonon density of state spectra obtained at 20 K *versus* energy transfer of ECF and Chol are presented in Figure 2 and 3, respectively. For hydrogenous substances, *G*(ν) function represents the vibrational density of states weighted by the amplitudes of hydrogen atoms vibrations in the given modes [25]. The resolution of the NERA spectrometer is close to 3% of the energy transfer in the energy transfer range below 800 cm$^{-1}$, then degrades to 5% in the region 800 to 1500 cm$^{-1}$ [27, 31]. The low temperature *G*(ν) spectra show a region of the lattice branch and the energy gap between the lattice and internal modes is ~130 cm$^{-1}$.

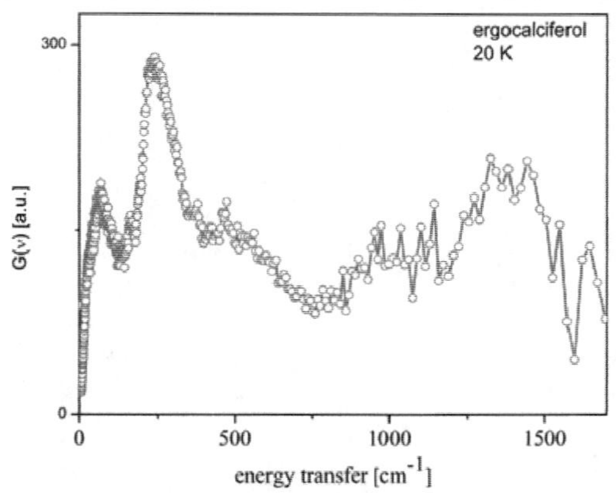


**Figure 2**. Generalized phonon density of state spectra of ECF at 20 K.

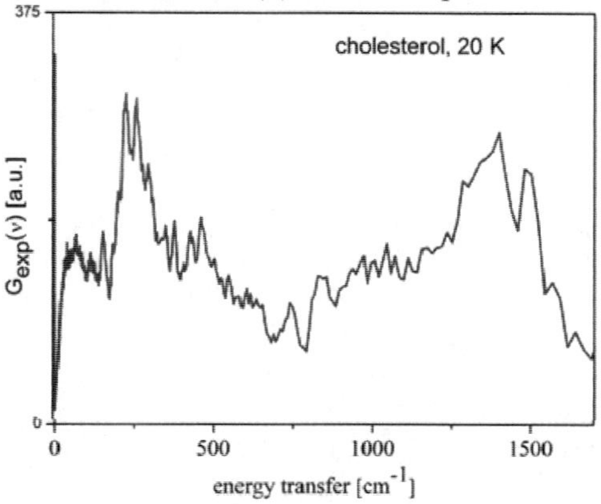


**Figure 3**. Generalized phonon density of state spectra of Chol at 20 K.

*3.2 Geometry optimisation*

Optimisation of the structure and calculation of the force field is necessary for determination of the vibrational spectrum. The X-ray studies [11] give the position of the skeletal atoms. The position of the hydrogen atoms is not known accurately, because of the low density of the electron cloud of hydrogen. The molecular structure of ECF was optimised by PM3 and DFT (B3LYP/6-311G**) calculations, and the bond length and internal angles are listed in Table 1, together with the data given from X-ray [11] studies.

The quality of the calculated parameters agreement with the experimental ones is expressed by the root mean square (RMS) deviation of the predicted parameters, defined as:

$$RMS = [\Sigma( X_{predicted} - X_{exp})^2/N]^{1/2} \quad (1)$$

where: $X_{predicted}$ and $X_{exp}$ can be either bond lengths or bond angles (in angstrom, degree) obtained after optimisation, and average bond lengths (of C–C, C=C, C–O type), bond angles (C–C–C, C–C–O, C–C=C) from the X-ray data [11], respectively; *N* is the number of bonds lengths, angles or normal modes of internal vibration. Only the structure of the carbon skeleton with attached side chain was used for comparisons as the hydrogen atom positions determined by the X-ray diffraction were insufficiently accurate. In the carbon skeleton one may find different types of C–C bonds, *i.e.* C($sp^3$)–C($sp^3$), C($sp^3$)–C$sp^2$), C($sp^2$)–C($sp^2$), C($sp^2$)=C($sp^2$), C($sp^3$)–OH h lying in the range (1.406–1.592), (1.433–1.542), (1.444–1.486), (1.303–1.388) and (1.406–1.447) (Å), respectively. The B3LYP method leads to optimization of the geometric parameters of the ECF skeleton which are close to the experimental data. It is seen from Table 1, that the most reliable bond length RMS values are provided by the B3LYP/ 6-311 G** calculations. As follows from Tables 1 and 2, the RMS for diffraction study (two molecules ), DFT and PM3 methods for C-C bond and C-C-C angles are:

**Table 1.** Collection of bond length (Å) and angles (deg) of ECF.

| bond | Subunit A [11] | Subunit B [11 ] | PM3 | B3LYP/ 6-311G** | angles | Subunit A [11] | Subunit B [11 ] | PM3 | B3LYP/ 6-311G** |
|---|---|---|---|---|---|---|---|---|---|
| C1-C2 | 1.534 | 1.527 | 1.58 | 1.59745 | C1-C2-C3 | 108.66 | 113.13 | 109.00 | 106.16240 |
| C2-C3 | 1.448 | 1.525 | 1.47 | 1.46161 | C2-C3-C4 | 111.42 | 108.29 | 114.35 | 119.88073 |
| C3-C4 | 1.504 | 1.542 | 1.49 | 1.44520 | C3-C4-C5 | 113.10 | 112.90 | 111.47 | 111.25527 |
| C4-C5 | 1.498 | 1.538 | 1.54 | 1.60732 | C4-C5-C10 | 111.57 | 113.21 | 115.24 | 118.54323 |
| C5-C10 | 1.468 | 1.443 | 1.42 | 1.38050 | C5-C10-C1 | 114.04 | 117.54 | 119.35 | 122.40723 |
| C10-C1 | 1.433 | 1.509 | 1.44 | 1.48854 | C2-C3-O | 107.80 | 110.99 | 116.30 | 119.94558 |
| C3-O | 1.446 | 1.406 | 1.37 | 1.35469 | C5-C10-C19 | 122.22 | 123.47 | 120.03 | 121.64809 |
| C10-C19 | 1.349 | 1.388 | 1.47 | 1.50688 | C10-C5=C6 | 126.41 | 125.24 | 128.05 | 130.03903 |
| C5=C6 | 1.348 | 1.305 | 1.35 | 1.39022 | C5=C6-C7 | 124.31 | 128.40 | 124.99 | 125.74461 |
| C6-C7 | 1.488 | 1.454 | 1.44 | 1.43336 | C6-C7=C8 | 125.04 | 126.60 | 122.86 | 127.79179 |
| C7=C8 | 1.322 | 1.346 | 1.34 | 1.35879 | C7-C8-C9 | 127.83 | 124.35 | 124.87 | 124.96392 |
| C8-C9 | 1.500 | 1.493 | 1.50 | 1.51249 | C8-C9-C11 | 112.59 | 109.49 | 109.18 | 112.41627 |
| C9-C11 | 1.478 | 1.535 | 1.53 | 1.54456 | C9-C11-C12 | 113.66 | 110.22 | 112.77 | 113.20114 |
| C11-C1 | 1.590 | 1.563 | 1.53 | 1.54045 | C11-C12-C13 | 109.29 | 109.49 | 109.59 | 112.28642 |
| C12-C13 | 1.515 | 1.529 | 1.53 | 1.54250 | C12-C13-C14 | 109.49 | 109.65 | 108.06 | 107.17500 |
| C13-C14 | 1.529 | 1.565 | 1.55 | 1.56885 | C13-C14-C8 | 110.61 | 107.34 | 108.06 | 113.23245 |
| C14-C8 | 1.530 | 1.520 | 1.49 | 1.50452 | C14-C8-C9 | 110.53 | 111.17 | 111.22 | 111.40457 |
| C13-C17 | 1.563 | 1.589 | 1.56 | 1.56334 | C13-C14-C15 | 105.84 | 104.98 | 104.29 | 104.23119 |
| C17-C16 | 1.525 | 1.566 | 1.55 | 1.56324 | C14-C15-C16 | 101.51 | 104.75 | 104.02 | 103.92314 |
| C16-C15 | 1.547 | 1.564 | 1.53 | 1.55118 | C15-C16-C17 | 107.97 | 106.29 | 106.65 | 107.30340 |
| C15-C14 | 1.522 | 1.493 | 1.52 | 1.53104 | C16-C17-C13 | 105.50 | 101.59 | 102.59 | 103.88699 |
| C13-C18 | 1.523 | 1.506 | 1.52 | 1.54166 | C17 –C19-C14 | 97.91 | 97.90 | 97.94 | 97.89000 |
| C17-C20 | 1.541 | 1.544 | 1.53 | 1.55169 | C12-C13-C18 | 112.00 | 110.59 | 108.93 | 110.73144 |
| C20-C21 | 1.516 | 1.572 | 1.52 | 1.54162 | C17-C20-C21 | 116.20 | 110.76 | 112.94 | 113.83790 |
| C20-C22 | 1.529 | 1.472 | 1.50 | 1.51166 | 13-C17-C20 | 119.84 | 121.60 | 119.88 | 119.51329 |
| C22=C23 | 1.350 | 1.375 | 1.33 | 1.33273 | C17-C20-C22 | 109,12 | 107.21 | 109.30 | 109.97003 |
| C23-C24 | 1.541 | 1.520 | 1.50 | 1.50937 | C20-C22=C23 | 124.48 | 121.32 | 122.65 | 125.99516 |
| C24-C25 | 1.506 | 1.550 | 1.54 | 1.50937 | C22=C23-C24 | 127.41 | 119.78 | 123.29 | 125.48607 |
| C24-C28 | 1.467 | 1.537 | 1.52 | 1.54112 | C23-C24-C25 | 110.22 | 113.50 | 110.41 | 113.48907 |
| C25-C26 | 1.488 | 1.543 | 1.52 | 1.53420 | C23-C24-C28 | 113.79 | 108.40 | 109.74 | 110.35568 |
| C25-C27 | 1.407 | 1.493 | 1.52 | 1.53635 | C24-C25-C26 | 111.19 | 110.94 | 110.80 | 113.93068 |
| | | | | | C24-C25-C27 | 116.56 | 110.42 | 111.85 | 111.43903 |
| | | | | | C26-C25-C27 | 110.07 | 111.94 | 111.18 | 110.57426 |
| ***RMS C=C*** | ***0.013*** | ***0.028*** | ***0.008*** | ***0.031*** | ***RMS C-C=C*** | ***1.29*** | ***3.03*** | ***2.29*** | ***3.12*** |
| ***RMS C-C*** | ***0.050*** | ***0.048*** | ***0.045*** | ***0.079*** | ***RMS C-C-C*** | ***5.68*** | ***5.72*** | ***5.62*** | ***6.06*** |
| | | | | | ***RMS C-C-$CH_3$*** | ***4.96*** | ***4.72*** | ***3.64*** | ***3.72*** |

**Table 2.** Collection of bond length (Å) and angles (deg) of Chol.

| bond | Subunit A [11] | Subunit B [11 ] | PM3 | B3LYP/ 6-311G** | angles | Subunit A [11] | Subunit B [11 ] | PM3 | B3LYP/ 6-311G** |
|---|---|---|---|---|---|---|---|---|---|
| C1-C2 | 1.544 | 1.539 | 1.52 | 1,53363 | C1-C2-C3 | 111.13 | 111.14 | 109.23 | 111.29251 |
| C2-C3 | 1.500 | 1.501 | 1.54 | 1,52705 | C2-C3-C4 | 109.43 | 109.01 | 109.94 | 110.23953 |
| C3-C4 | 1.533 | 1.494 | 1.53 | 1,52552 | C3-C4-C5 | 113.50 | 113.76 | 111.95 | 112.52614 |
| C4-C5 | 1.488 | 1.532 | 1.50 | 1,51468 | C4-C5-C10 | 117.60 | 116.96 | 117.20 | 116.85591 |
| C5-C10 | 1.543 | 1.504 | 1.51 | 1,53483 | C5-C10-C1 | 106.38 | 109.70 | 109.37 | 108.26074 |

| | | | | | | | | | |
|---|---|---|---|---|---|---|---|---|---|
| C10-C1 | 1.554 | 1.546 | 1.54 | 1,55844 | C10-C1-C2 | 113.51 | 114.69 | 111.70 | 114.57716 |
| C3-O | 1.425 | 1.420 | 1.42 | 1,42827 | C2-C3-O | 112.50 | 112.70 | 111.78 | 112.27340 |
| C10-C19 | 1.539 | 1.551 | 1.53 | 1,54899 | C5-C10-C19 | 109.01 | 108.76 | 197.60 | 108.67031 |
| C5=C6 | 1.322 | 1.333 | 1.34 | 1,33543 | C10-C5=C6 | 121.18 | 121.68 | 122.74 | 122.7565 |
| C6-C7 | 1.519 | 1.498 | 1.48 | 1,50012 | C5=C6-C7 | 125.66 | 125.86 | 124.42 | 125.33216 |
| C7-C8 | 1.518 | 1.557 | 1.53 | 1,53473 | C6-C7-C8 | 112.45 | 113.91 | 113.18 | 113.17235 |
| C8-C9 | 1.536 | 1.511 | 1.54 | 1,54942 | C7-C8-C9 | 109.05 | 109.92 | 109.77 | 109.77776 |
| C9-C10 | 1.535 | 1.543 | 1.55 | 1,56661 | C8-C9-C10 | 112.26 | 114.65 | 110.74 | 112.60443 |
| C9-C11 | 1.511 | 1.543 | 1.53 | 1,5442 | C9-C5-C10 | 110.24 | 110.80 | 110.5 | 110.19153 |
| | | | | | C8-C9-C11 | 111.05 | 110.32 | 112.55 | 111.98614 |
| C11-C12 | 1.548 | 1.509 | 1.52 | 1,54037 | C9-C11-C12 | 115.53 | 111.23 | 113.96 | 114.27051 |
| C12-C13 | 1.544 | 1.515 | 1.53 | 1,53996 | C11-C12-C13 | 108.81 | 107.84 | 110.72 | 111.93147 |
| C13-C14 | 1.512 | 1.542 | 1.55 | 1,55322 | C12-C13-C14 | 107.35 | 114.02 | 106,72 | 106.37701 |
| C14-C8 | 1.527 | 1.492 | 1.53 | 1,53226 | C13-C14-C8 | 113.90 | 109.93 | 111.34 | 115.45453 |
| C13-C17 | 1.592 | 1.552 | 1.55 | 1,56769 | C14-C8-C9 | 109.40 | 110.32 | 108.29 | 109.89053 |
| C17-C16 | 1.557 | 1.546 | 1.55 | 1,56577 | C13-C14-C15 | 104.54 | 105.21 | 103.77 | 104.29025 |
| C16-C15 | 1.549 | 1.553 | 1.53 | 1,55103 | C14-C15-C16 | 103.39 | 102.16 | 104.46 | 103.67630 |
| C15-C14 | 1.526 | 1.528 | 1.52 | 1,53343 | C15-C16-C17 | 107.29 | 108.25 | 107.53 | 107.26520 |
| C13-C18 | 1.545 | 1.517 | 1.53 | 1,54322 | C16-C17-C13 | 102.47 | 102.77 | 103.14 | 103.32287 |
| C17-C20 | 1.516 | 1.555 | 1.53 | 1,54894 | C17-C13-C14 | 100.28 | 99.71 | 100.67 | 100.05856 |
| C20-C21 | 1.526 | 1.535 | 1.52 | 1,53662 | C12-C13-C18 | 110.08 | 111.74 | 110.26 | 110.82050 |
| C20-C22 | 1.534 | 1.552 | 1.53 | 1,54759 | C17-C20-C21 | 112.53 | 109.19 | 111.22 | 113.17761 |
| C2-C23 | 1.546 | 1.528 | 1.52 | 1,53518 | C13-C17-C20 | 120.71 | 121.20 | 116.44 | 119.63869 |
| C23-C24 | 1.560 | 1.527 | 1.52 | 1,53398 | C17-C20-C22 | 111.63 | 111.47 | 111.10 | 110.33884 |
| C24-C25 | 1.542 | 1.471 | 1.53 | 1,54034 | C20-C22-C23 | 114.33 | 113.56 | 110.20 | 115.30068 |
| C25-C26 | 1.492 | 1.492 | 1.52 | 1,53539 | C22-C23-C24 | 107.80 | 110.61 | 111.94 | 112.60153 |
| C25-C27 | 1.560 | 1.405 | 1.52 | 1,5346 | C23-C24-C25 | 106.26 | 115.98 | 111.31 | 115.48069 |
| | | | | | C24-C25-C26 | 94.08 | 101.24 | 110.44 | 110.46618 |
| | | | | | C24-C25-C27 | 84.42 | 118.25 | 110.54 | 112.66006 |
| | | | | | C26-C25-C27 | 118.76 | 109.92 | 110.51 | 110.46548 |
| ***RMS C-C*** | ***0.023*** | ***0.033*** | ***0.015*** | ***0.019*** | ***RMS C-C-C*** | ***6.921*** | ***4.768*** | ***3.424*** | ***4.251*** |
| ***RMS C=C*** | ***0.005*** | ***0.005*** | ***0.212*** | ***0.008*** | ***RMS C-C=C*** | ***2.246*** | ***2.097*** | ***0.840*** | ***1.364*** |
| | | | | | ***RMS C-C-CH₃*** | ***12.031*** | ***5.597*** | ***2.995*** | ***4.003*** |
| | | | | | ***RMS C-C=O*** | ***0.1*** | ***0.1*** | ***0.82*** | ***0.327*** |

([0.050, 0.048, 0.079, 0.044) Å, (5.6 °, 5.7 °, 6.0 °, 5.6°)] and [(0.0226, 0.0325, 0.0148, 0.0079) Å, (6.9°, 4.7°, 3.4°, 4.2°)] for ECF and Chol, respectively. The most reliable bond lengths values are given by the DFT/B3LYP/6-311G** calculations. The observed agreement between optimised structures of an isolated molecule (by the DFT/B3LYP/6-311G** method) with the X-Ray data may indicates only weak intermolecular interactions.

In order to verify the optimised structure of the studied molecular systems, the mid-infrared (IR) spectra were recorded at room temperature and compared with that calculated by the quantum chemistry methods. Figures 3 and 4 show the experimental and calculated by DFT/B3LYP/6-311G*** method spectra for ECF and Chol, respectively. The need for a scaling factor testifies to the harmonic interactions.

### *3.3. Calculations / G(v)*

In the experimental $G(\nu)$ spectra of ECF an energy gap appears at ~130 cm$^{-1}$ (Figure 6). It separates the lattice branch and internal modes. In the range 130-300 cm$^{-1}$, characteristic of the torsional out-of-plane modes of methyl groups, the spectrum is smeared. In the previously determined neutron vibration spectra of other steroids in which the four-ring steroid skeleton was substituted by only two

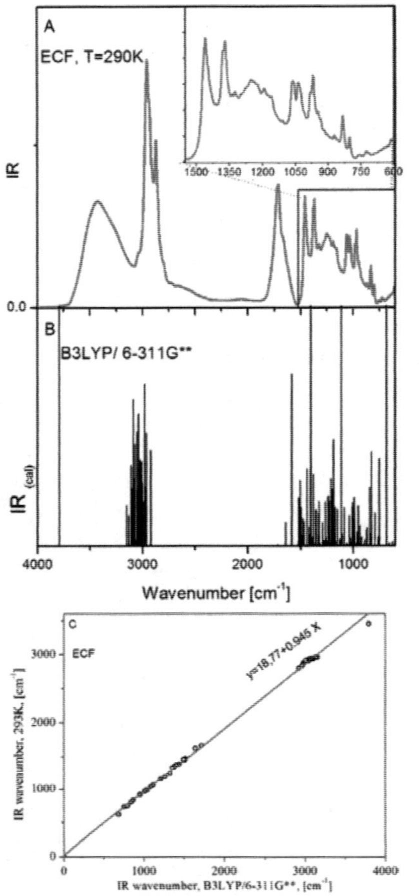


**Figure 4.** Comparison of experimental FT IR spectra (a) with calculated ones by DFT method for ECF (b); Collection of experimental and calculated frequencies of normal modes (c).

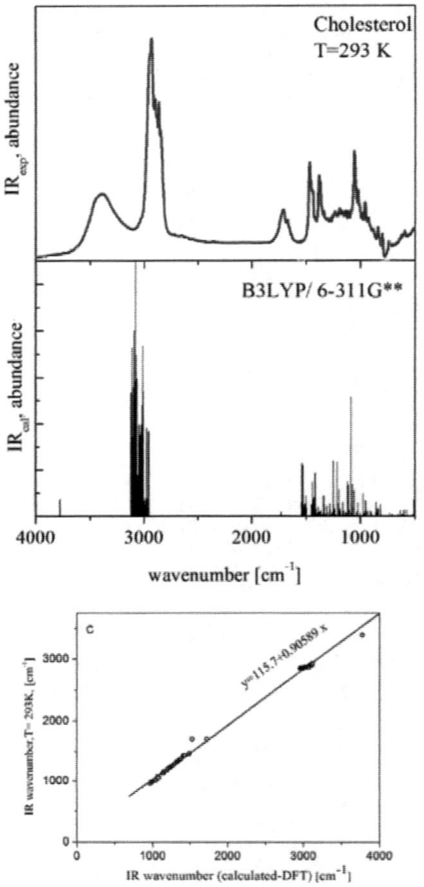


**Figure 5.** Comparison of experimental FT IR spectra (a) with calculated ones by DFT method for Chol (b); Collection of experimental and calculated frequencies of normal modes (c).

or three methyl groups, the torsional out-of-plane modes of subsequent methyl groups were clearly visible [12-17].

The question arises if the smearing of this fragment of the spectrum is related to the effect of the methyl groups from the side chain. Therefore the phonon density of state spectra of Chol was recorded at 20 K. It was compared with the calculated $G_{cal}(\nu)$, in Figure 7. The calculations were performed in the isolated molecule approximation. The harmonic vibrational wavenumbers of normal modes computed by quantum chemistry (QC) methods and the corresponding experimental values are collected in Table 3 and 4.

Table 5 compares the frequencies of the torsional out-of-plane modes of subsequent methyl groups (with maximum PED) together with the barrier heights to their reorientation, both in ECF and Chol. On the basis of the B3LYP/6-311G** calculations, the bands corresponding to the torsional out-of-plane vibrations of methyl groups should appear in the range 160-260 and 140-240 $cm^{-1}$ in the spectra of ECF and Chol, respectively. The band corresponding to the torsional out-of-plane vibrations of $C(18)H_3$ and $C(19)H_3$ should occur at 176, 211 and 176, 188, 220, $cm^{-1}$ in the spectrum of ECF and at 145, 212 and 194, 205, 212 $cm^{-1}$ in that of Chol.

Journal of Physics: Conference Series **554** (2014) 012010 doi:10.1088/1742-6596/554/1/012010

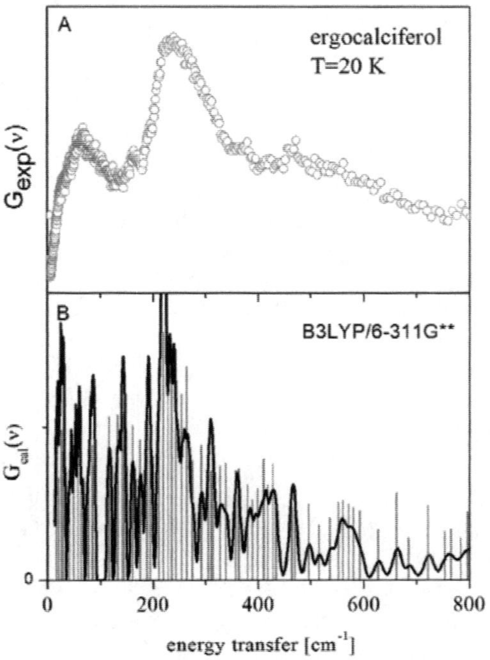


**Figure 6.** Comparison of the low temperature phonon density of states $G_{exp}(\nu)$ spectra (A) of ECF with $G^{cal}(\nu)$ calculated in the single molecule approximation by DFT/B3LYP/6-311G** method. (B).

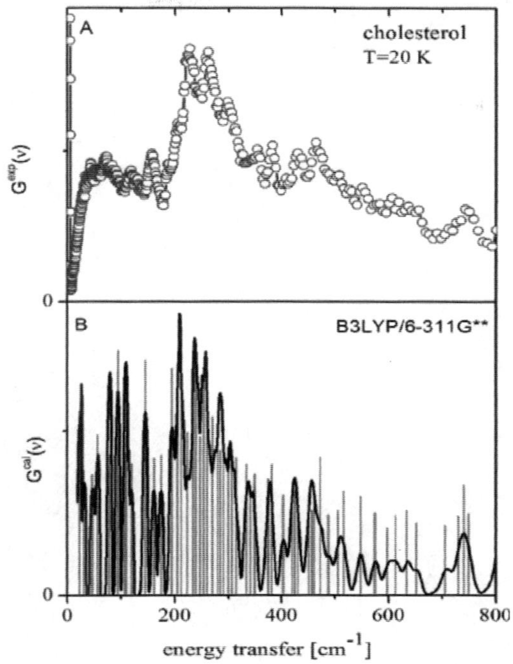


**Figure 7.** Comparison of the low temperature phonon density of states $G^{exp}(\nu)$ spectra (A) of Chol with $G^{cal}(\nu)$ calculated in the single molecule approximation by DFT/B3LYP/6-311G** method. (B).

The calculated frequency and intensity of each mode is given by a δ function (red bar) which after convolution with the resolution function of the spectrometer gives the solid black line [31].

**Table 3.** Experimental frequencies of normal modes from IINS (20 K) and IR (290 K) spectroscopy together with the calculated ones, for ECF.

| G(ν) 20 K [cm$^{-1}$] | IR 290K [cm$^{-1}$] | PM3 | B3LYP/6-311G** | Proposed assignment |
|---|---|---|---|---|
| | | | 16.86 | χ[C20-C22] |
| | | | 21.21 | χ[C7- C8] |
| | | | 26.72 | χ[C5-C6] , χ[C23-C24] |
| | | | 29.83 | χ[C20-C22] |
| | | 42.0 | 41.47 | χ [C20-C22] , χ [C5-C6] |
| | | 45.9 | 49.53 | χ [C4-C5] , χ [C1-C2] |
| | | 55.7 | 57.53 | χ [C17-C20] |
| 61.6±5 | | 75.7 | 63.92 | χ [C4-C5], χ [C20-C22] |
| | | 76.6 | 77.27 | χ [C23-C24] χ [C17-C20] |
| | | 97.6 | 84.99 | χ [C4-C5] , χ [C6-C7] |
| | | 110.3 | 86.87 | χ [C1-C10] |
| 106.5±5 | | 112.7 | 116.64 | χ [C1-C10] , χ [C3-C4] |
| | | 144.6 | 132.53 | χ [C13-C17] , χ [C16-C17] |
| | | 150.4 | 141.37 | χ [C12-C20] |

| | | | | |
|---|---|---|---|---|
| | | 167.7 | 144.5 | χ [C23-C24] |
| 166.2±5 | | 177.3 | 161.82 | χ [C24-C28] |
| | | 189.5 | 176.30 | χ [C10-C19], χ [C13-C18] |
| | | 194.9 | 188.96 | χ [C10-C19] |
| 193.9±5 | | 202.3 | 195.13 | χ [C20-C21]χ[C20-C22]χ [C10-C19] |
| | | 213.4 | 211.56 | χ [C10-C19] χ [C20-C21] 13-18 |
| | | 217.4 | 220.28 | χ [C25-C26] χ [C25-C27] χ [C10-C19] χ [C20-C21] |
| | | 219.8 | 220.56 | χ [C10-C19] χ [C25-C26] χ [C25-C27] χ [C24-C28] |
| | | 227.5 | 229.10 | χ [C25-C26] χ [C25-C27] χ [C24-C28] |
| 240.9±5 | | 233.2 | 236.37 | χ [C25-C26] χ [C25-C27] χ [C24-C28] χ [C20-C21] |
| | | 247.5 | 243.82 | χ [C25-C26]χ [C25-C27]χ [C24-C28] χ [C13-C18] |
| | | 263.7 | 254.83 | χ [C13-C18]χ [C20-C21] |
| *266±5* | | 267.8 | 264.87 | χ [C13-C18]χ [C20-C21 |
| | | 294.0 | 274.91 | χ [C25-C26]χ [C25-C27] χ [C24-C28] χ[C20-C21] χ [C13-C18] |
| 309±5 | | 308.6 | 292.08 | δ [C17-C20-C21] ρ [C10-C5-C4] |
| | | 312.0 | 305.06 | χ [C20-C21] ρ [C10-C5-C4] |
| | | 328.2 | 310.57 | δ [C21-C20-C22] |
| | | 333.1 | 314.39 | χ [C3-O29] χ [C5-C6] χ [C7-C8] χ [C2-C3] |
| | | 337.0 | 327.61 | χ [C3-O29] χ [C2-C3] |
| | | 343.8 | 338.78 | χ [C5-C6] χ [C7-C8] δ [C24-C24-H] χ [C3-O29] |
| | | 353.6 | 357.27 | δ [C24-C24-H49] δ [C24-C25-C26] χ [C20-C21] χ [C5-C6] |
| | | 382.5 | 363.42 | δ [C26-C25-C27] δ [C12-C13-C18] |
| 370±5 | | 386.2 | 380.23 | χ [C20-C21] δ [C2-C3-C4] χ [C2-C3] δ [C2-C3-O29] |
| | | 393.6 | 389.32 | χ [C20-C21] δ [C12-C13-C18] |
| | | 406.1 | 399.26 | χ [C1-C2] δ [C2-C3-C4] χ [C4-C5] |
| | | 422.5 | 410.37 | δ [C26-C25-C27] χ [C3-C4] χ [C4-C5] |
| 422±10 | | 430.2 | 419.14 | δ [C4-C3-O29] χ [C3-C4] |
| | | 446.3 | 429.00 | δ [C15-C14-C8] δ [C12-C13-C18] |
| | | 462.0 | 434.85 | δ [C18-C13-C17]δ [C21-C20-C22] |
| | | 467.3 | 464.70 | δ [C4-C3-O29] ρ [C14-C8-C7] |
| 468±10 | | 494.6 | 465.52 | δ [C20-C22-C24] δ [C12-C13-C17] δ [C24-C24-H49] |
| | | 500.4 | 470.82 | δ [C4-C5-C6] δ C10-C1-C2] δ [C2-C3-O29] |
| | | 503.1 | 496.51 | δ [C25-C24-C28] δ [C21-C20-C22] δ [C24-C25-C26] |
| | | 520.6 | 515.73 | δ [C4-C3-O29] δ [C2-C3-O29] δ [C3-C4-C5] |
| | | 544.6 | 536.36 | δ [C21-C20-C22] δ [C24-C24-H] δ [C24-C24-C25] χ [C22-C24] |
| 545±10 | | 559.4 | 553.05 | δ [C2-C3-O29] δ [C1-C2-C3] δ [C10-C5-C6] δ [C10-C5-C4] |
| | | 580.9 | 561.11 | δ [C13-C17-C20] δ [C17-C20-C21] |
| | | 604.8 | 572.10 | δ [C24-C25-C27] δ [C24-C24-C25] |
| | | 610.5 | 581.00 | ρ [C10-C5-C4] χ [C5-C6] |
| | | 622.7 | 592.72 | ρ [C10-C5-C4] δ [C7-C8-C9] |
| 616±20 | 613.26 | 627.2 | 627.06 | δ [C18-C13-C17] ρ [C14-C8-C7] |
| 660±20 | | 659.6 | 661.75 | δ [C14-C15-C16] δ [C15-C16-C17] |
| | | 701.0 | 685.29 | ρ [C14-C8-C7] δ [C13-C14-C8] |
| 705±20 | 727.0 | 721.5 | 722.04 | δ [C3-C4-C5] δ [C2-C3-C4] δ [C1-C2-C3] |
| | 744.4 | 774.8 | 753.03 | δ [C15-C16-C17] δ [C15-C16-C17] υ [C12-C13] δ [C16-C15-H] |
| | | 807.8 | 765.39 | υ [C1-C2] δ [C10-C1-C2] δ [C10-C1-H] |
| | | 831.3 | 783.92 | ρ [C5-C6-C7] υ [C1-C2] υ [C1-C10] δ [C1-C2-H] δ [C2-C1-H] |
| | 802.1 | 851.5 | 797.21 | υ [C24-C25] δ [C24-C28-H] |
| | | | 817.30 | υ [C13-C17] δ [C12-C11-H] |
| | | | 826.00 | ρ [C5-C6-C7]ρ [C6-C7-C8] χ [C6-C7] |

**Table 4.** Experimental frequencies of normal modes from IINS (20 K) and IR (290 K) spectroscopy together with the calculated ones, for Chol.

| G(ν)$^{exp}$ 20 K cm$^{-1}$ | IR $^{exp}$ 290 K cm$^{-1}$ | PM3 cm$^{-1}$ | Calculated frequencies of normal modes B3LYP/6 - 311G** cm$^{-1}$ | Proposed assignment |
|---|---|---|---|---|
| | | | 22.18 | χ [ C20-C22] |
| | | 51.42 | 30.38 | χ [ C9-C11] χ[ C20-C22] |
| 43.3±5 | | 73.36 | 46.27 | χ [ C9-C10] χ[ C23-C24] |
| | | 85.08 | 56.02 | χ [ C23-C24] χ[ C10-C5] |
| 70.4±5 | | 90.52 | 77.73 | χ [ 22-C23]χ[ C2-C3] |
| | | 102.28 | 78.47 | χ [C17-C20] χ[ 22-C23] |
| | | 111.34 | 93.61 | χ [ C23-C24] |
| | | 120.89 | 106.74 | χ [C17-C20] χ [C24-C25] |
| | | 136.30 | 109.92 | χ [C20-C22] χ [ C23-C24] |
| | | 162.61 | 118.69 | χ [ 22-C23] |
| 118.7±5 | | 174.30 | 145.00 | χ [C18-C13]χ[C16-C17] |
| 158.4±5 | | 179.97 | 161.20 | χ [C17-C20], χ[C22-C20], |
| 186.0±5 | | 205.60 | 174.87 | χ [C20-C22] χ[C18-C13] |
| | | 212.71 | 194.34 | χ [C10-C19] χ[C18-C13] |
| | | 219.58 | 205.26 | χ [C18-C13] χ[C20-C21] |
| 206.3±5 | | 229.12 | 212.22 | χ [C18-C13] χ[C10-C19] χ [C20-C21] |
| 225.2±5 | | 237.82 | 222.72 | χ [C10-C19] χ[C20-C21] |
| | | 244.38 | 233.54 | χ [C25-C27] χ[C25-C26] |
| | | 248.83 | 239.24 | χ [C25-C27] χ [C25-C26] |
| | | 250.25 | 246.62 | χ [C20-C21] χ[C18-C13 |
| | | 259.03 | 253.70 | χ [C10-C19] |
| 260.0±5 | | 270.47 | 260.03 | χ [C25-C27] χ[C25-C26] |
| | | 278.00 | 270.10 | χ [C25-C27] χ[C25-C26] χ [C18-C13] |
| | | 289.58 | 279.66 | χ [C1-C2] χ[C7-C8] |
| | | 306.37 | 284.06 | χ [C3-C2] χ[C20-C22] |
| | | 310.94 | 290.21 | χ [C7-C8] χ[O1-C3] |
| 297.5±5 | | 332.10 | 299.56 | χ [O1-C3] χ[C1-C10] δ[C1-C10-C19] |
| | | 344.87 | 306.13 | δ [C1-C10-C19] δ[ C9-C10-C19] |
| | | 363.10 | 314.11 | δ[O-C3-C4] δ[O-C3-C2] |
| | | 382.80 | 333.12 | δ [C18-C13-C14] δ[C21-C20-C22] |
| | | 386.96 | 341.01 | δ [C21-C20-C22] δ[C24-C25-C27] δ[O-C3-C2] |
| 350.7±5 | | 401.82 | 348.68 | δ [C24-C25-C26] δ[C12-C13-C18] δ[C26-C55-C27] |
| | | 411.74 | 374.03 | δ [C17-C20-C21] δ[C12-C13-C18] |
| 381.9±5 | | 420.13 | 380.63 | δ [C24-C25-C27] δ[C26-C25-C27] δ [C17-C13-C18] |
| | | 436.50 | 403.01 | δ [C26-C25-C27] δ[C17-C20-C22] |
| | | 448.32 | 419.79 | δ [C12-C13-C18] δ[ C9-C10-C19] |
| 428.4±10 | | 459.08 | 424.60 | δ [C24-C25-C27] δ [C21-C20-C22]δ [C17-C20-C21] |
| | | 481.58 | 430.73 | δ [C9-C8-C14] δ[C11-C12-C13] |
| | | 482.35 | 451.30 | δ [C24-C25-C27] δ[C21-C20-C22] δ [C17-C20-C21] |
| | | 492.34 | 457.05 | δ [C9-C10-C19] δ[C8-C9-C11] |
| 464.5±15 | | 496.23 | 460.92 | δ [C26-C5-C27] δ[C12-C13-C18] δ [C20-C22-C23] |
| | | 511.78 | 472.82 | δ [C2-C3-C4] δ[O-C3-C2] |

| | | | | |
|---|---|---|---|---|
| | | 527.06 | 487.77 | δ [C6-C5-C10] δ[C6-C7-C8] |
| 507.4±15 | | 530.89 | 505.23 | δ [O-C3-C2] δ[C1-C2-C3] |
| | | 553.90 | 516.06 | δ [C21-C20-C22] δ[C20-C22-C23] |
| 552.1±15 | | 582.24 | 547.51 | δ [C8-C9-C11] δ[C13-C17-C20] |
| | 576.6 | 607.21 | 574.49 | δ [C13-C17-C20] δ[C17-C20-C21] |
| | 591.1 | 613.37 | 597.88 | ρ [C4-C5-C6] δ[C5-C10-C19] δ [C13-C17-C20] |
| 609.0±20 | | 636.92 | 612.82 | ρ [C4-C5-C6] δ[C1-C10-C5] |
| 638.3±20 | 624.2 | 655.30 | 633.83 | ρ [C4-C5-C6] δ[C5-C10-C19] |
| | 642.2 | 676.88 | 652.69 | δ[C18-C13-C17] |
| | 667 | | | δ[C13-C17-C16] |
| | 700 | 722.33 | 706.15 | δ[C5-C6-C7] δ[C6-C7-C8] |
| | | 773.18 | 730.43 | χ [22-C23] χ[C23-C24] δ[C22-C23-H] |
| | 739 | 778.28 | 739.83 | δ[C15-C16-H] δ[H-C16-C17] |
| 746.0±20 | | 799.23 | 749.22 | ρ [C5-C6-H] δ[C6-C7-H] |
| | 800 | 823.58 | 804.43 | ρ [C5-C6-H] υ[C10-C19] |
| 841.5±20 | | 839.09 | 810.93 | δ [H-C1-C2] υ[C2-C3] |

**Table 5**. Position of torsional out-of-plane modes χ[C-$CH_3$] (with maximum PED) collected with the height of the barrier of $C_3$ methyl group rotation of Chol and ECF predicted by DFT/B3LYP/6-311G**, respectively.

| Torsional out-of-plane mode | Predicted position of methyl torsion by DFT of Chol [$cm^{-1}$] | barrier high for methyl rotation [kcal/mol] of Chol | Predicted position of methyl torsion by DFT of ECF [$cm^{-1}$] | barrier high for methyl rotation [kcal/mol] of ECF |
|---|---|---|---|---|
| C-C(18)$H_3$ | 145, 212 | 0.75 | 176, 211 | 0.82 |
| C-C(19)$H_3$ | 194, 212 | 0.82 | 176, 220 | 0.48 |
| C-C(21)$H_3$ | 205, 212, 222 | 2.21 | 195, 236, 264 | 1.71 |
| C-C(26)$H_3$ | 233, 239 | 1.69 | 220, 229, 236 | 1.64 |
| C-C(27)$H_3$ | 233, 239 | 1.86 | 220, 236, 248 | 1.89 |
| C-C(28)$H_3$ | | | 168, 246, 274 | 2.51 |

The torsional vibrations of the methyl group of the side chain C(26)$H_3$ and C(27)$H_3$ should be at 220, 229, 236 and 220, 236, 248 $cm^{-1}$ in the spectrum of ECF and at 233, 239 and 233, 239 $cm^{-1}$ in that of Chol, respectively. The out-of-plane torsions of C(21)$H_3$ are predicted at 195, 236, 264 in ECF while at 205, 212, 222 $cm^{-1}$ in Chol, respectively. In the spectra of both ECF and Chol the out-of-plane C–C torsion modes of ring A and D appear close to 42-92 and 30-110 $cm^{-1}$, respectively (*i.e.* in the lattice branch).

The torsional out-of-plane modes of the methyl groups were predicted to have different values of the energy transfer, which suggests dynamic inequivalence of the methyl groups. The methyl groups attached to the four-ring carbon skeleton are more mobile in Chol, then those attached to the three-ring of ECF. Moreover, the position of the out-of-plane torsion of the methyl group attached to the side chain was predicted at higher energy transfer, than those attached to the skeleton, in both analysed compounds.

The in-plane deformation vibrations assigned to δ[CCO] were calculated to appear at 202 $cm^{-1}$ in the spectrum of ECF while at 290 $cm^{-1}$ in the spectrum of Chol. The deformational in-plane modes δ[CCC(26)] and δ[CCC(28)] of the terminal methyl groups were predicted to occur at 290 and 353, 382 $cm^{-1}$ in ECF, and in Chol at 247 and 290, 310 $cm^{-1}$, respectively. The deformational in-plane modes δ[CCC] of ring A are at 386, 504 $cm^{-1}$ and between 470-500 $cm^{-1}$ for ECF and Chol, respectively.

### *3.4.[1]H NMR*

The suggestion arising from the DFT calculations of the vibrational spectra, of different transition energies of the torsional out-of-plane modes of methyl groups of steroids, as well as different values of thermal displacement coefficient given by X-ray study, have stimulated further study of the internal dynamics by $^1$H NMR. To describe the shape of $^1$H NMR absorption spectrum of ECF, which depends on temperature, it is obvious to analyse the second moment and slope line width of each signal. Figure 8 presents the temperature dependencies of the second moment of $^1$H NMR line of ECF. At 50 K the second moment $M_2$ is 23.5 $\times \cdot 10^{-8}$ T$^2$. On heating it decreased to 12.5 $\times \cdot 10^{-8}$ T$^2$ at 200 K.

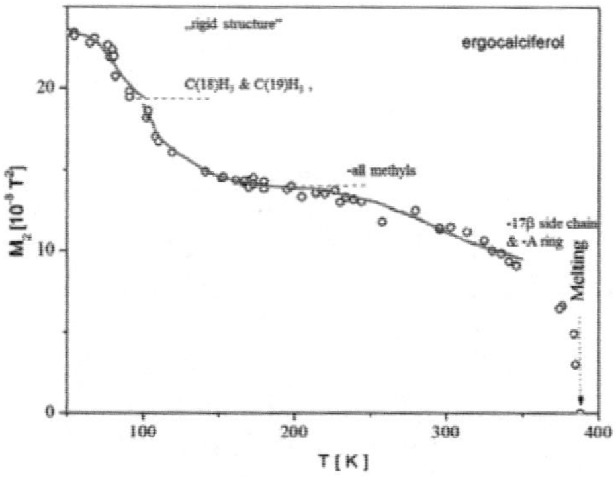


**Figure 8.** Temperature dependence of the second moment of $^1$H NMR line of EFC (open circle).

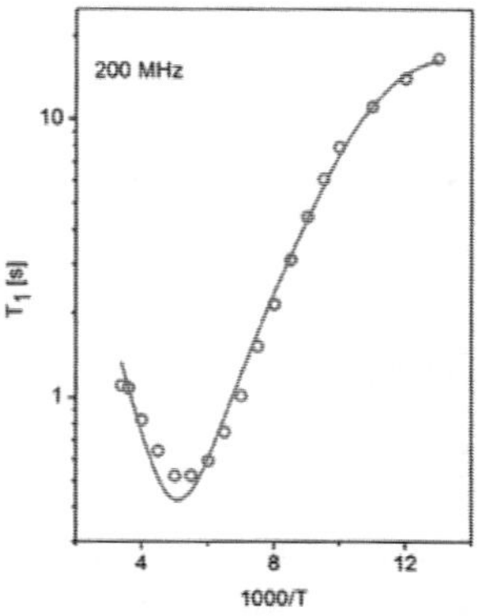


**Figure 9.** Spin-lattice relaxation time versus reciprocal temperature (open circle) of EFC at 200 MHz.

Continuous lines - fitted $M_2(T)$ (eq.3) and $T_1(1/T)$ (eq.4) for following internal motion: rotation of $C(18)H_3$ and $C(19)H_3$ with $E_a$= 0.9 kcal/mol, next remaining four methyls with $E_a$= 1.4 kcal/mol, respectively.

The experimental value of $M_2$ obtained above by the graphical integration of the $^1$H NMR line can be discussed against the values obtained from the crystal structure data, according to the formula given by van Vleck [34]:

$$M_2^{II} = \frac{3}{5}\gamma_I^{\,2}\hbar^2 I(I+1)\frac{1}{N}\sum_{I,S}^{N}\frac{1}{r_{IS}^6} \qquad (2)$$

where: $\gamma_I$ is the gyromagnetic coefficient of the resonant nucleus spin *I*, $r_{IS}$ is the proton-proton distance. $M_2$ was calculated from the formula given by van Vleck [34] in two steps as the intra and intermolecular contributions to $M_2$. Also the effects of various molecular reorientations are summarized elsewhere [18-22, 35, 36, 37, 38]. Given the structural parameters from the X-ray study of Hull [11], and the data obtained after optimisation by quantum chemistry DFT calculation, it is possible to calculate the intramolecular part of the second moment $M_2$ for the rigid structure $M_2^{rigid,\text{int}ra}$ as 21.71$\cdot\times$ 10$^{-8}$ T$^2$. Taking into account the crystal structure [11], the intermolecular H–H contribution to $M_2$ [33] is estimated as 1.7 $\times\cdot 10^{-8}$ T$^2$.

On heating, the onset of the following reorientations should be taken into account:

-methyl group reorientation around the threefold symmetry axis of the C–C bond,

-the *trans-gauche* isomerisation of the 17β side chain,

-the reorientation of side chain around the C(24)–C(25) bond and ring A in ECF.

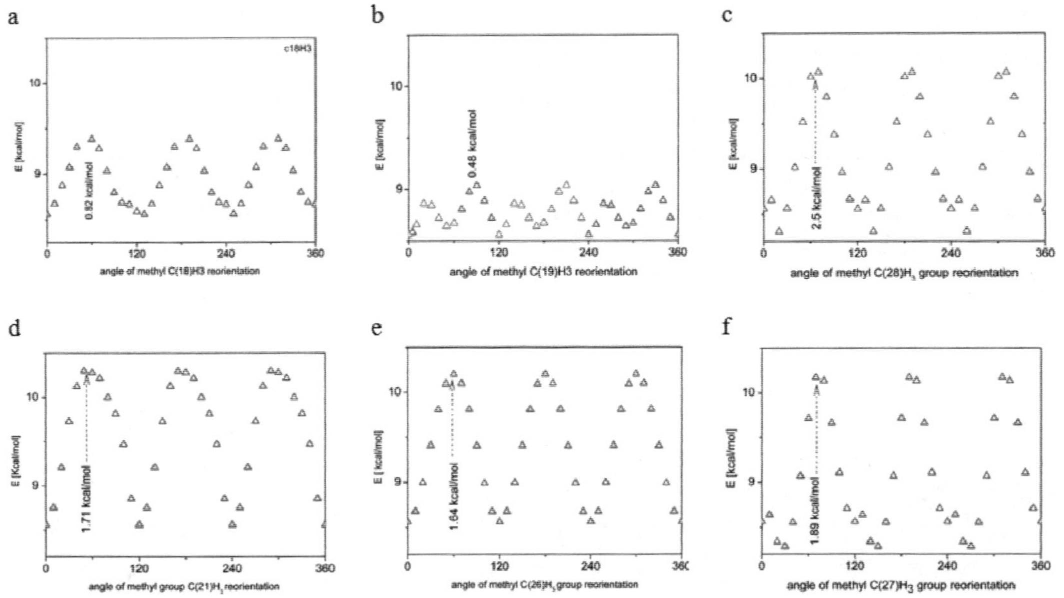


**Figure 10.** Energy of isolated molecule versus position of subsequent methyl groups of EFC internal rotation of the methyl groups $C(18)H_3$ and $C(19)H_3$ attached to the four–ring structure are 0.82 and 0.48 kcal/mol, respectively, while those of $C(21)H_3$, $C(28)H_3$, $C(27)H_3$, $C(26)H_3$

In order to propose the sequence of onset of the methyl group reorientation of ECF, the energy of the isolated molecule *versus* the orientation of the methyl group about the three-fold symmetry axis of the C–C bond were calculated by the PM3 method in the isolated molecule approximation. The heights of the three-fold symmetry barrier for methyl reorientation are presented in Figure 10. It allows us to conclude about the sequence of the onset of methyl group reorientation. The heights of the barriers for reorientations are as follows: 1.71, 2.5, 1.64, 1.89 kcal/mol. The activation energies of rotation of methyl groups which are attached to the side chain are higher than those for the rotation of the methyl attached to the steroid skeleton. Thus, the cause for the decrease in $M_2$ from $21.5 \cdot \times 10^{-8}$ T$^2$ to $12.5 \cdot \times 10^{-8}$ T$^2$ and then to $8.0 \cdot \times 10^{-8}$ T$^2$ for ECF is the subsequent reorientation of two and four methyl groups about their $C_3$ axis, respectively. *Trans-gauche* isomerisation of the 17β-side chain leads to a decrease in $M_2$ by $1.0 \cdot \times 10^{-8}$T$^2$. Next, reorientation of the 17β side chain of ECF about the C(23)–C(24) bond and ring A motion cause a reduction of $M_2$ to $6.0 \cdot \times 10^{-8}$ T$^2$.

The change in $M_2$ versus temperature caused by the internal dynamics is observed when the internal reorientation takes place with a frequency of a few kHz (*i.e.* it is of the order of the linewidth). According to the simple formula for the temperature dependence of the second moment $M_2$ given by Gutowsky and Pake [36]:

$$M_2 = \left(M_2^{reo}\right) + \left(\left(M_2^{rigid}\right) - \left(M_2^{reo}\right)\right) \frac{2}{\pi} arctg(\alpha \gamma_I \sqrt{M_2} / 2\pi \nu_c) \qquad (3)$$

where $M_2^{rigid}$ and $M_2^{reo}$ are the values of the second moment of $^1$H NMR line above and below the narrowing region, respectively, $\nu_c$ is the Arrhenius correlation frequency ($\nu_c = \nu_{co}\, e^{-E_a/RT}$), one may calculate the activation parameters for the internal motion of subsequent molecular groups. The activation energy of two and four methyl group reorientation around the $C_3$ symmetry axis was estimated as (0.9± 0.2) and (1.4 ± 0.3) kcal /mol, respectively. The barrier height of the side-chain groups motion is (4.0 ± 0.4) kcal /mol. Figure 8 collects the experimental data of $M_2$ vs $T$, together with the fitted values of $M_2$ according to formula (4) and mentioned above activation energies.

Figure 9 presents the temperature dependence of the spin-lattice relaxation time for the studied compound. The magnetisation recovery was single-exponential in the whole temperature range within the experimental accuracy of 3%. The temperature curve of $T_1$ shows a minimum of 51 ms close to 196 K, and asymmetrical branches with a smaller slope on the low-temperature side. When we consider the reorientation of a few methyl groups, the spin-lattice relaxation rate is [37, 38]:

$$\left(\frac{1}{T_1}\right) = \sum_i C_i \left( \frac{\tau_{c,i}}{1+\omega_o^2 \tau_{c,i}^2} + \frac{4\tau_{c,i}}{1+4\omega_o^2 \tau_{c,i}^2} \right) \tag{4}$$

where: $C_i$ is the relaxation constant, $\tau_{ci}$ is the correlation time of methyl group reorientation ( $\tau_{ci}=1/\nu_{ci}$), $\omega_o$ is a resonance frequency.

We estimated the distribution of the activation energy of methyl group reorientation as 2:4 with average activation energies 0.9 and 1.4 kcal/mol, respectively, the correlation time $\tau_{01} \sim 10^{-13}$ s. As shown in Figure 8, the model satisfactorily describes the experimental data. The sequence of onset of internal reorientations was in qualitative agreement with the ones predicted from the calculations performed in the isolated molecule approximation.

The study of internal dynamics for different steroids has been previously performed by the proton magnetic resonance method, among others for Chol [18-24]. The data obtained from those earlier studies permitted Andrew and Peplinska [24] to conclude that there was a distribution of the activation energies of the methyl groups, when they are attached to steroid skeleton at different positions, because the temperature dependence of spin-lattice relaxation times was asymmetric and indicated different slopes on their right and left branch. The authors concluded that there were two relaxation processes [24]. The weaker one is attributed to the reorientation of one methyl group about the $C_3$ axis, and the stronger relaxation process - to the reorientation of the remaining four methyl groups. Thus the distribution of relaxation times may find an explanation in the values of the activation energies calculated in the isolated molecule approximation, and presented in Figure 9. The calculations of the total energy of isolated molecule of Chol versus orientation of methyl groups denoted subsequently as $C(18)H_3$, $C(19)H_3$ and $C(21)H_3$, $C(26)H_3$, $C(27)H_3$ (attached to the four-ring skeleton and β-side chain, respectively), give the height of the three-fold symmetry barrier as: 0.75, 1.61 and 2.21, 1.69, 1.85 kcal/mol, respectively. Therefore, the methyl group $C(18)H_3$ (substituent to the steroid skeleton) is more mobile than the remaining four groups in the molecule. This confirms the distribution of activation energies observed by Andrew and Peplinska[24].

## 4. Concluding remarks

- Calculations of the energy of the isolated molecule *vs.* position of protons for methyl groups have revealed different heights of the barriers of $C_3$ reorientations, which is lower for the skeleton methyl groups $C(18)H_3$ and $C(19)H_3$ than for those attached to 17β side chain of ECF as well as of Chol.
-In particular, the calculated value of the activation energy of the methyl group denoted as $C(18)H_3$ of Chol is lower than the remaining four groups in the molecule, and the torsional out-of-plane mode $\chi$[C-C(18)H$_3$ is predicted at lower energy transfer in G(ν) spectra. Distribution of the relaxation process takes place.
-Positions of the torsional-out-of-plane modes of methyl groups have been predicted by the DFT calculations. The position of the torsional–out-of-plane modes of $CH_3$ groups attached to the steroid skeleton were calculated at lower value of energy transfer than those of the torsional-out-of-plane modes of methyl groups attached to the 17β side chain.
-The dependencies $M_2(T)$ as well as $T_1(1/T)$ of $^1$H NMR signals of ECF have proven the dynamical inequivalence of the methyl groups. The ratio of the number of the fastest to the slowest methyl groups was 2:4, which corresponds to the numbers of methyl group attached to the ring structure: named -$C(18)H_3$ and -$C(19)H_3$, and the four of the side chain: -$C(26)H_3$, -$C(27)H_3$, -$C(28)H_3$, -$C(21)H_3$, respectively.

**5. Acknowledgement**

The QC calculations were performed at PSCC in Poznan.
Financial support under the grant of Polish Plenipotentiary at JINR in Dubna, is gratefully acknowledgment by the authors.

**6. References**

[1] Shienh H-S, Hoard L G, Nordman C E, 1981 *Acta Cryst.* **B37** 1538-1543
[2] Hsu L-Y, Kampf J W, Nordman C E, 2001 *Acta Cryst.* **B58** 260-264
[3] Hull S E, Woolfsen M M, 1976 *Acta Cryst.* **B32** 2370-2373
[4] Knobler C, Romer C, Braun P B, Hornstra J, 1972 *Acta Cryst.* **B28** 2097-2103
[5] Lobkovsky E B, *Private communication* 2000, after 2013 CCDC v 3.3
[6] Mercury CSD 3.3, Macrae C F, Bruno J A, Chisholm J A, Edgington P R, McCabe P, Pidcock E, Rodriguez-Monge L, Taylor R, van de Streek J, Wood P A, 2008 *J. Appl. Cryst.* **41** 466.
[7] Mercury, Macrae C F, Edgington P R, McCabe P, Pidcock E, Shields G P, Taylor R, Towler M, van de Streek J, 2006 *J. Appl. Cryst.* **39-3** 453
[8] Bruno I J, Cole J C, Edgington P R, Kessler M K, Macrae C F, McCabe P, Pearson J, Taylor R, 2002 *Acta Cryst.* **B58** 389
[9] Taylor R, Macrae C F, 2001 *Acta Cryst.* **B57** 815
[10] The Cambridge Structural Database, Allen F H, 2002 *Acta Cryst.* **B58** 380
ConQuest, Bruno I J, Cole J.C., Edgington P R, Kessler M, Macrae C F, McCabe P, Pearson J, Taylor R, 2002 *Acta Cryst.* **B58** 389
[11] Hull S E, Leban I, Main P, White P S, Woolfson M M, 1976 *Acta Cryst.* **B32** 2374
[12] Szyczewski A, Holderna-Natkaniec K, Natkaniec I, 2004 *J. Mol. S*truct. **693** 49
[13] Szyczewski A, Holderna-Natkaniec K, Natkaniec I, 2003 *J. Mol. Struct.* **650** 69
[14] Holderna-Natkaniec K, Natkaniec I, Nowak D, Jurga K, Szyczewski A, 2005 *Chem. Phys.* **317** 178
[15] Hołderna-Natkaniec K, Natkaniec I, Jurga K, Nowak D, Szyczewski A, 2006 *Solid State Phenomena* **112** 93
[16] Szyczewski A, Hołderna-Natkaniec K, Natkaniec I, 2004 *J. Mol. Struct.* **698** 41
[17] Hołderna-Natkaniec K, Szyczewski A, Natkaniec I, Khavryutchenko V, Pawlukojc A, 2002 *Appl.Phys. A – Mat. Sci. Process* **74** [Suppl 1] S1274
[18] Andrew E R, Jurga K, Radomski J M, Reynhardt E C, 1992 *Solid State NMR* **1** 121
[19] Andrew E R. Kempka M, 1993 *Solid State NMR.* **2** 261
[20] Andrew E R, Kempka M, Szyczewski A, 1996 *Molecular Physics* **88** 605
[21] Szyczewski A, Hołderna-Natkaniec K, Natkaniec I, 2004 *Solid State NMR* **25** 21
[22] Andrew E R, Radomski J, 1993 *Solid State NMR* **2** 57
[23] Szyczewski A, Hołderna-Natkaniec K, 2005 *J. Mol.Struct.* 7**43** 129
[24] Andrew E R, Peplinska B, 1990 *Molecular Physics* **70** 505
[25] Lovesey S W, Thermal neutron Scattering, Academic Press, London, New York, 1965 & G. L. Squires, *Introduction to the Theory of the Thermal neutron Scattering*, Cambridge University Press, London, 1997.
[26] *Neutron Data Booklet*, editors Dianoux A J, Lander G, 2003 ILL Neutron for Sciences, Grenoble
[27] Natkaniec I, Bragin S I, Brankowski J, Mayer J, 1994 in *Proceedings of the ICANS XII Meeting*, Abington 1993, RAL Report 94-025, vol.**I.**, p.89
[28] Gaussian 09, Revision **A.1**, Frisch M J, Trucks G W, Schlegel H B, Scuseria G E, Robb M A, Cheeseman J. R., Scalmani G., Barone V., Mennucci B., Petersson G. A., Nakatsuji H., Caricato M, Li X, Hratchian H P, Izmaylov A F, Bloino J, Zheng G, Sonnenberg J L, Hada M, Ehara M, Toyota K, Fukuda R, Hasegawa J, Ishida M, Nakajima T, Honda Y, Kitao O, Nakai H, Vreven T, Montgomery J A, Jr, Peralta J E, Ogliaro F, Bearpark M, Heyd J, Brothers E, Kudin K N, Staroverov V N, Kobayashi R, Normand J, Raghavachari K, Rendell A, Burant J C, Iyengar S S, Tomasi J, Cossi M, Rega N, Millam J M, Klene M, Knox J E, Cross J B, Bakken V, Adamo C,

Jaramillo J, Gomperts R, Stratmann R E, Yazyev O, Austin A J, Cammi R, Pomelli C, Ochterski J W, Martin R L, Morokuma K, Zakrzewski V G, Voth G A, Salvador P, Dannenberg J J, Dapprich S, Daniels A D, Farkas Ö, Foresman J B, Ortiz J V, Cioslowski J, and Fox D J, Gaussian, Inc., Wallingford CT, 2009.

[29] Stewart J, 1989 J. *Comput. Chem.* **10** (2) 209
Stewart J, 1989 *J. Comput. Chem.* **10** (2) 221
Stewart J, 2004 *J. Mol. Modeling* **10** (2) 155

[30] Khavryutchenko V D, 1990 COSPECO Complex PROGRAM for Vibrational spectroscopy, Institute of Surface Chemistry Ukrainian Academy of Science, Kiev,

[31] Kazimirov W J, Natkaniec I, 2003 Programme for Calculation of the Resolution Function of NERA-PR and KDSOG-M Inelastic Neutron Scattering Inverse Geometry Spectrometers, Preprint **P14-2003-48** JINR Dubna

[32] Ramirez-Cuest A J, 2004 *Comp. Phys. Comm.* **157** 226

[33] TOSCA at www.isis.rl.ac.uk

[34] van Vleck J H, 1948 *Phys. Rev.* **74** 1168

[35] Smith G W, 1969 J. *Chem. Phys.* **51** 3569

[36] Gutowsky H S, Pake G E, 1950 *J .Chem. Phys.* **18** 162

[37] Bloembergen N, Purcell E M, Pound R V, 1948 *Phys. Rev.* **73** 679

[38] Kubo R, Tomita K, 1954 *J. Phys. Soc. Japan.* **9** 888

Dynamics of Molecules and Materials-II IOP Publishing
Journal of Physics: Conference Series **554** (2014) 012011 doi:10.1088/1742-6596/554/1/012011

# pH-induced size changes in solutions of cholesteric liquid-crystal polymers studied by SANS

**Mercedes Pérez Méndez[1], Daniel Rodriguez Martinez[2,3] and Stephen M King[3].**
[1] Institute of Polymer Science and Technology, Group of Physical-Chemistry and Modelization of Macromolecules (PCMM), C/ Juan de la Cierva, 3. 28006 Madrid, Spain
[2] Retinal degeneration: from genetics to therapy Laboratory, Centro Andaluz de Biología Molecular y Medicina Regenerativa Edif. CABIMER Avda. Américo Vespucio, s/n Parque Científico y Tecnológico Cartuja 93. 41092 Sevilla, Spain
[3] ISIS Facility, STFC Rutherford Appleton Laboratory, Harwell Oxford, Didcot, Oxfordshire OX11 OQX, United Kingdom

E-mail: perezmendez@ictp.csic.es

**Abstract.** Polycations possessing substantial buffering capacity below physiological pH, are intrinsically efficient transfection agents. These vectors have been shown to deliver genes as well as oligonucleotides, both *in vitro* and *in vivo*, by protecting DNA from inactivation by blood components. Their efficiency relies on extensive endosome swelling and rupture that provides an escape mechanism for the polycation/DNA complexes. Recently, biocompatible cationic cholesteric liquid-crystal polymers (ChLCP) have proved able to condense and successfully transfect DNA, acting as non-viral vectors. Here the radius of gyration of the new ChLCPs is determined by SANS as a function of pH, the ultimate aim being to correlate changes in polymer conformation with membrane activity. With increasing pH the polymers apparent radii of gyration increased to a maximum, before subsequently decreasing. This molecular expansion, on passing from acidic pH environment (*cf.*, lysosome pH 3.5- 4, late endosome pH 5- 6, early endosome pH 6- 6.5) to neutral pH (cytosol pH=7-7.4), matches the endocytic route through the cell, where the pH change is used as a signal to release biomacromolecules, such as DNA. It confirms that the new cationic ChLCPs could act as an endosomolytic release system in gene therapy according to the hypothesis of "the proton sponge".

## 1. Introduction

The introduction of exogenous genetic material in cells is a key stage in the development of basic research in cellular biology. The term "transfection" is used to indicate the transfer of DNA into the nuclei of cells of higher organisms [1]. The development of *non-viral* DNA delivery vectors make it possible to model viral assembly and gene transfer by incorporating functional groups that enable particular assembly and transfer steps.

Cationic polymers (at physiological pH) are already used to condense anionic nucleic acids into nano-sized particle-like complexes called "polyplexes", through self-assembly driven by electrostatic interactions. By compressing DNA molecules to a relatively small size cellular internalization is facilitated and, thus, transfection efficacy is improved [2]. Whilst the intravascular route of administration is an attractive approach for wide spread delivery, it is however particularly plagued by toxicity and biodistribution problems [3, 4]. New types of delivery vector are always welcome.

Several non-permanent polycations possessing substantial buffering capacity below physiological pH, such as lipopolyamines and polyethylenimines, are intrinsically efficient transfection agents, i.e. they do not need the addition of lysosomotropic bases, or cell targeting, or membrane disruption agents. These vectors have been shown to deliver genes as well as oligonucleotides both *in vitro* and *in vivo*. Their efficiency relies on extensive endosome swelling and rupture that provides an escape mechanism for the polycation/DNA polyplexes [5, 6]. The challenge, therefore, is to see if materials chemistry can provide more efficient, and 'smart', delivery vectors. One promising area of research uses liquid crystals (LCs).

Liquid crystals are self-organizing systems that form intermediate mesophases, combining the order of perfect crystals with the mobility of fluids [7]. These mesophases may be formed either by the effect of temperature (thermotropic) or of solvent concentration (lyotropic). Lyotropic liquid crystals, in the form of lipids, are of course ubiquitous throughout the life sciences [8]. They are a prerequisite for the development of life and the ability of cells to function.

In all LCs the longitudinal molecular axes are aligned parallel to each other, with the distribution of the molecular centres determining the type of LC. The most common types are the nematic (centres distributed isotropically), figure 1a, and smectic (where they are arranged in layers), figure 1b. Less common are the cholesteric, figure 1c, where nematic ordering is stacked in a superhelical structure, with a certain pitch, figure1d. It is these LCs that are the focus of this paper.

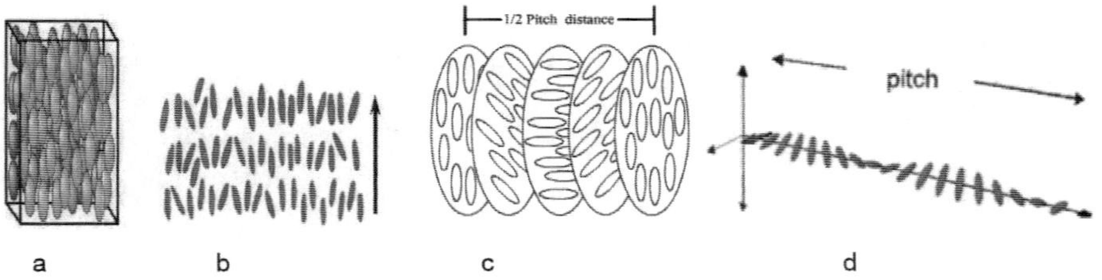


**Figure 1.** Liquid Crystal mesophases. (a) Nematic, (b) Smectic, (c, d) Cholesteric

Cholesteric liquid-crystal polymers (ChLCP), designated PTOBDME $[C_{34}H_{36}O_8]_n$ with the monomeric chemical formula shown in figure 2, and PTOBEE $[C_{26}H_{20}O_8]_n$ having the same mesogenic rigid group along the main chain but different side chain length, have been synthesised in our laboratory, through a stereoselective polycondensation reaction [9, 10]. The amphiphilic nature of the monomers makes them polymerize into helical chains.

These ChLCP have been extensively characterized [11]. They are both thermotropic and lyotropic, and adopt different conformations when dispersed in solution, depending on the solvent nature and concentration. They are also optically active and biocompatible against macrophages and fibroblast cellular lines, making them truly multifunctional materials.

We have previously shown that due to their lyotropic behaviour, these ChLCPs can entrap smaller molecules inside [12] and interact with liquid-crystalline biomacromolecules, such as nucleic acids and lipids (both neutral and cationic). The structures of the complexes formed have been previously studied by simultaneous synchrotron SAXS/WAXS [13, 14, 15].

Subsequently, formulations of the synthetic ChLCPs and cationic monomeric surfactants were developed in our lab to entrap an anionic plasmid of DNA. These complexes were successfully tested as non-viral vectors in gene therapy, both *in vitro* and *in vivo* [16, 17].

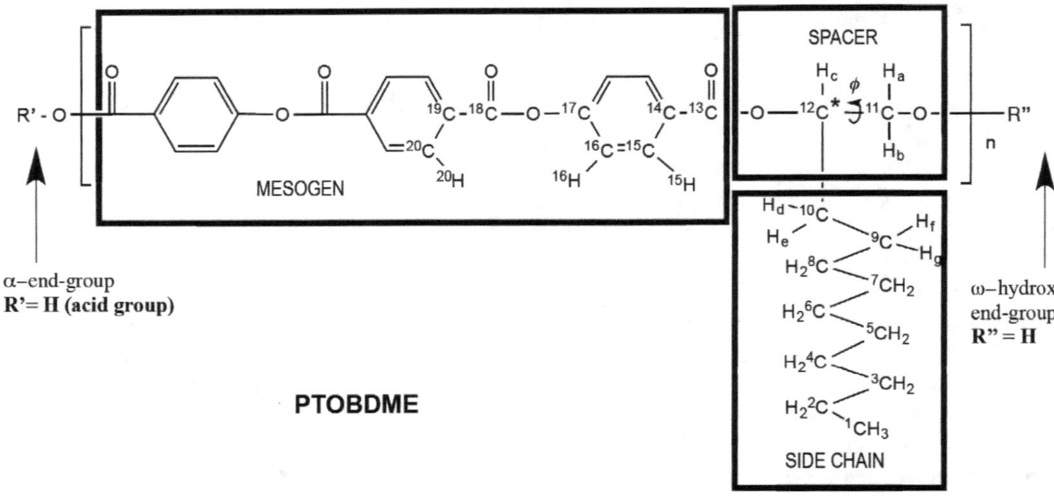


**Figure 2.** Chemical formula of PTOBDME $[C_{34}H_{36}O_8]_n$ subdivided into mesogen, spacer and side chain. PTOBEE $[C_{26}H_{20}O_8]_n$ shares the same mesogen and spacer but only has two carbon atoms in the side chain.

In order to specifically interact with complex polynucleotides, new *cationic* ChLCPs were synthesized, by functionalization of the precursors PTOBEE and PTOBDME with nitrogenated groups ($NH_2$, choline, etc) [18]. The molecular model of Poly[PTOBEE]$_{10}$-$NH_2$ is shown in figure 3.

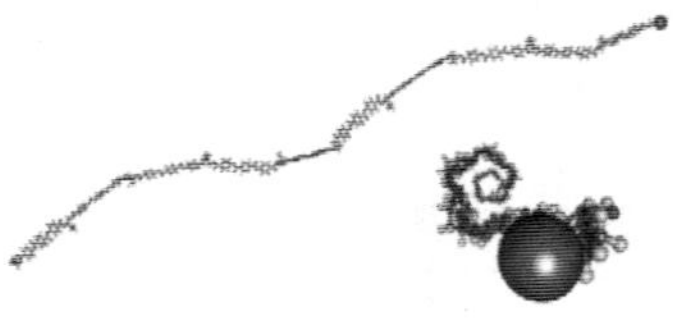

**Figure 3.** Molecular model of Poly[PTOBEE]$_{10}$-$NH_2$.
In the lower right sectional perspective view the terminal N atom of the amine end group is represented by a ball.

Polyplexes of the cationic ChLCP PTOBEE-$NH_2$ with the commercial anionic polynucleotide [Poly-C-Poly-G] have been studied by contrast variation small-angle neutron scattering (SANS), exploiting the difference in neutron scattering length densities (NSLD) of the two components (1.887E+10/cm$^2$ for PTOBEE-$NH_2$, 3.320E+10/cm$^2$ for [Poly-C-Poly-G]) to separately distinguish both component structures, and to highlight the whole polyplex, without any selective deuteration [19]. This provided complementary information to that obtained by SAXS [20].

In this paper SANS is used without contrast variation to investigate the pH-dependent conformational behaviour of three cationic ChLCPs. The pH value is a critical environmental parameter during transfection as it can be used as a signal to promote the efficient delivery of biomacromolecules or DNA to the correct intracellular compartment. The pH of an endosome is lower than that of the cytosol by up to two pH units, depending on the stage of endosomal development.

## 2. Experimental

Three cationic ChLCPs have been studied: PTOBDME-choline, PTOBDME-$NH_2$, and PTOBEE-$NH_2$. All samples were dispersed in aliquots of $D_2O$ (to maximise the scattering contrast), pH-adjusted with

5M HCL or 5M NaOH (to match the conditions encountered during endocytic internalization; pH 3 to 7.4), at a fixed polymer concentration of 7 mg/ml. The pH's of the solutions were re-checked after dissolution of the polymers. In adjusting the pH, allowance was made for the small offset between pH and pD: pD=pH+0.4.

Contrast variation (via adjusting $D_2O/H_2O$ ratio) was not employed in this work, since the neutron scattering length densities (NSLD) of all the ChLCPs studied (between 1.3 - 3.3 ($x10^{10}$ $cm^{-2}$)), are significantly different to that of $D_2O$ (+6.3) to provide excellent contrast. As the SANS signal is proportional to the square of the difference in NSLD between the solute and solvent, using solvent mixtures with $H_2O$ (-0.5 on the same scale) would have merely reduced the contrast. Contrast variation would, however, provide a means to elucidate individual contributions to the overall scattering in the polyplexes in future work.

SANS data were obtained on the LOQ small-angle diffractometer at the ISIS Pulsed Neutron Source (STFC Rutherford Appleton Laboratory, Didcot, U.K.) [21a, 21b]. This is a fixed-geometry "white beam" time-of-flight instrument which utilizes neutrons with wavelengths, λ, between 2 and 10 Å. Data are simultaneously recorded on two, two-dimensional, position-sensitive, neutron detectors, to provide a simultaneous Q ($=4\pi/\lambda.\sin(\theta/2)$) range of 0.008–1.6 $Å^{-1}$, where θ is the scattering angle. Each polymer sample and corresponding solvent background was placed in 1 mm path length quartz disc-shaped cuvettes (volume 0.25 ml) and was measured for a total of 2 h in order to gather data of sufficient statistical precision. Each raw scattering data set was then corrected for the detector efficiencies, sample transmission and background scattering and converted to scattering cross-section data (denoted here as I vs Q) using the instrument-specific software [21c, 21d]. These data were placed on an absolute scale ($cm^{-1}$) using the scattering from a standard sample (a solid blend of hydrogenous and perdeuterated polystyrene) in accordance with established procedures [21e]. The data from the ChLCPs were analysed with a combination of partial Zimm plots (to extract the apparent radius-of-gyration of the nanostructure) [22] and Log-Log plots (*aka* Swollen Coil plots, to determine the fractal scaling of the nanostructure). The Zimm plots were fitted over the range $0.008 \leq Q \leq 0.02$ $Å^{1}$, whilst the Log-Log plots were fitted over a slightly extended range $0.008 \leq Q \leq 0.04$ $Å^{-1}$.

**3. Results and Discussion.**

Figure 4 shows the small-angle neutron scattering (SANS) patterns, in log I(Q) versus Q representation, from cationic ChLCPs PTOBDME-Choline, PTOBDME-$NH_2$ and PTOBEE-$NH_2$ dispersed in $D_2O$ at 7 mg/ml, as a function of pH (= 1.3, 3.58, 5.58, 7.2). These pH values were chosen to match the values in different cellular compartments (lysosome between pH 3.5 - 4, late endosome pH 5 - 6, early endosome pH 6 - 6.5 to cytosol, pH=7 - 7.4). Though superficially similar, there are some marked differences in both the intensity and Q-dependence of the scattering from the polymers as the pH changes, indicating that the size and/or shape of the solvated species is changing.

A zeroth-order approach to quantifying these changes is simply to determine the radius-of-gyration, $R_g$, of the ChLCPs from the SANS data. Traditionally this can be done through one of two low-angle linearisations, the Guinier plot or the Zimm plot [22]. We have used a variation of the latter as it has a slightly wider range of applicability (ie, is a little more robust to the presence of polydispersity). Whilst light scatterers will be familiar with the Zimm plot as a complex double extrapolation to infinite dilution and Q=0 (to correct for the strong angular dependence of the scattering), this is not actually necessary in order to estimate an apparent $R_g$ as the slope of the Zimm plot, 1/I(Q) vs $Q^2$, is just ($R_g^2/3$) x intercept as $Q < 0$, and SANS is always considered point scattering because the neutron wavelength is much larger than the object from which it is being scattered (the nucleus). The SANS data in Figure 4 are replotted in this partial Zimm representation (to denote that extrapolation to zero concentration was not performed) in Figure 5. Linearity is expected when $Q.R_g < \pi$ (note: this is a slightly broader limit [23] than Zimm's original definition, which allows for the presence of some polydispersity). For $R_g$'s up to ~200 Å this would imply $0 < Q^2 < 0.00025$ $Å^{-2}$. In some instances linearity would seem to extend beyond this range, in others a slight curve to the Zimm plot is evident

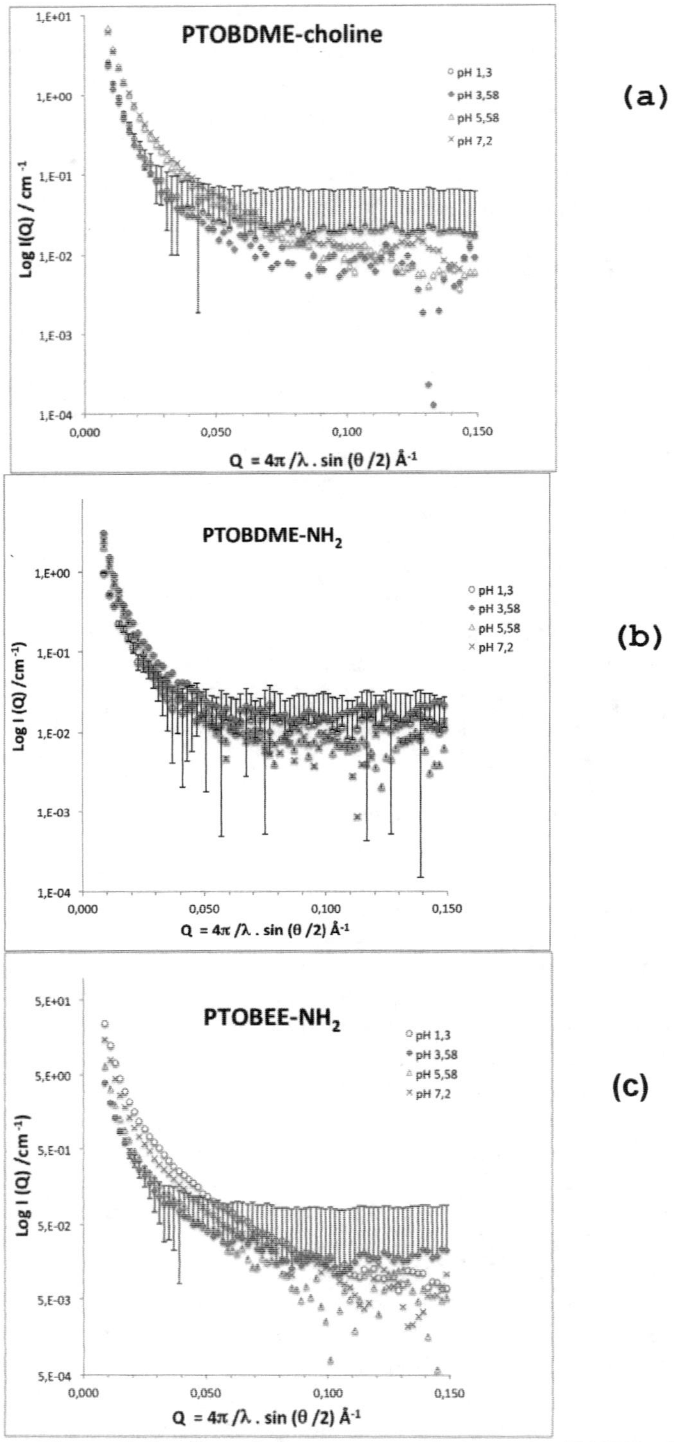


**Figure 4.** Small-angle neutron scattering patterns (SANS) of cationic ChLCPs, dispersed in $D_2O$ at 7 mg/ml, as a function of pH:
(a) PTOBDME-Choline; (b) PTOBDME-$NH_2$; (c) PTOBEE-$NH_2$.

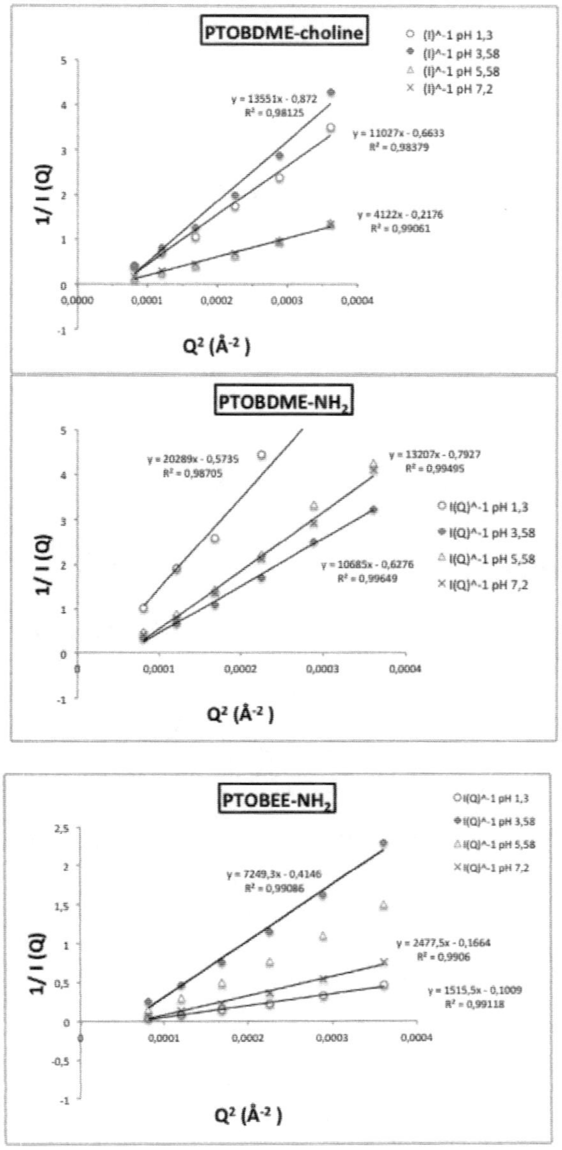


(a)

(b)

(c)

**Figure 5.** Partial Zimm plots of the cationic ChLCPs, dispersed in $D_2O$ at 7 mg/ml, as a function of pH: (a) PTOBDME-Choline; (b) PTOBDME-$NH_2$; (c) PTOBEE-$NH_2$. The slope is proportional to $R_g^2/3$ at small Q.

suggesting that a more limited range might be more appropriate. But for consistency we report $R_g$'s determined over the same range of $Q^2$.

In Figure 6 the apparent z-average radii of gyration, $R_g$, of the polymers extracted from the partial Zimm plots are displayed as a function of pH. With increasing pH, the apparent $R_g$ of all the polymers increases up to a maximum value, before subsequently decreasing.

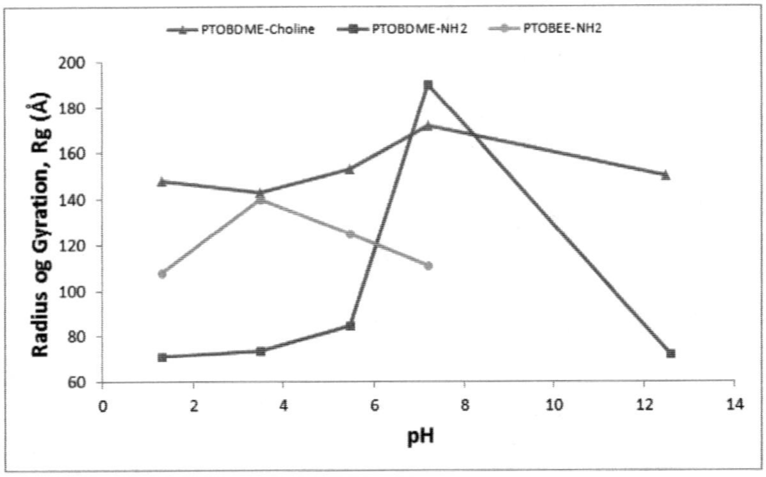


**Figure 6.** Variation of the apparent radius of gyration (Å) of the cationic polymers dispersed in $D_2O$ at [7 mg/ml] derived from the SANS data as a function of pH: (triangles) PTOBDME-Choline (squares) PTOBDME-$NH_2$ and (circles) PTOBEE-$NH_2$.

The cholesteric liquid crystalline nature of these new cationic polymers means they might reasonably be expected to have 'less random coil' solution conformations than other synthetic polymers, and there would seem to be evidence of this. In figure 7, the swollen coil fits to the SANS data from PTOBDME-$NH_2$ at pH 1.3 and pH 7.2 are shown. The gradients of the two linear regions are both significantly larger than what would be expected for an extended (filamentous) nanostructure, at -1.00, a 'self-avoiding chain' (ie, in a good solvent environment), at -1.66, or a 'Gaussian chain' (ie, in a theta solvent environment), at -2.00. Indeed the scaling displayed is more typical of some sort of network structure. The change observed in fractal scaling is also modest despite a large change in the apparent $R_g$ (almost a factor of 3 for this polymer).

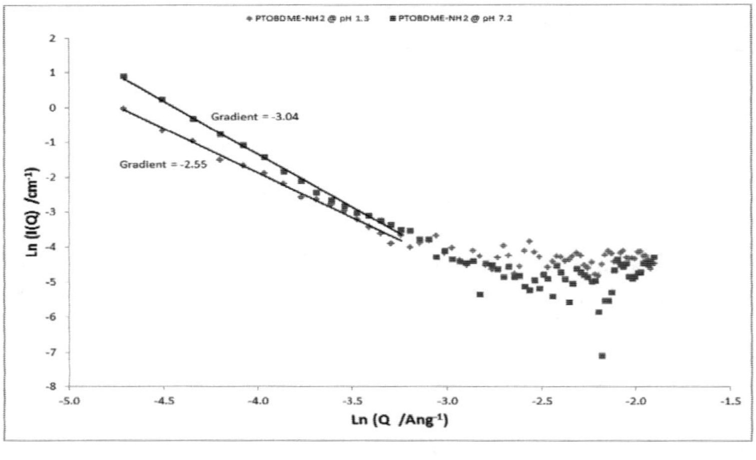


**Figure 7.** Swollen coil fits to the SANS from PTOBDME-NH2 at pH 1.3 and pH 7.2. The gradients of the straight lines are -2.55 and -3.04, respectively.

The next best level of analysis would be to try and fit the SANS data to analytical descriptions of different shaped objects. We have tried this, using the well-known model-fitting program, FISH [28]. However we were unable to obtain a fit in which we had both confidence and belief that it represented physical reality. As an example, in Figure 8 fits of the SANS from PTOBDME-$NH_2$ at pH= 1,3 and 7,2 against a Debye Gaussian coil model are shown [29]. Neither calculation is a good description of the data; a not altogether surprising fact given the fractal scaling seen in Figure 7 but it illustrates the difficulties. The solution architecture of the ChLCPs is evidently quite challenging to determine. The analysis most likely to elucidate the structure is perhaps a MD simulation *constrained* against the experimental SANS data. We note that software packages to achieve this, explicitly accounting for solvent, are starting to become available [30].

The apparently large pH-induced change in apparent $R_g$ (almost a factor of 3) for these ChLCP, on passing from acidic pH environment (*cf.* lysosome, endosome), towards neutral pH (*cf.* cytosol) and the posterior decrease in the more basic extreme, could be explained by the hypothesis of a molecular expansion up to a maximum, similar to the coil expansion described for other polymers at the same pH values [24], [25], [26]. The effect of extreme protonation/ deprotonation should be considered on intramolecular H-bridges (N-H...O) and on possible cleavage through ester groups, due to charge change.

The cleavage mechanism of a polyplex carrier, for delivery of a plasmid DNA, had been reported through S-S bonds when passing from pH 7 to 5.5 [27].

We have previously reported a decrease in the $R_g$ of the precursor PTOBEE in both $D_2O$ and in ethanol-$d_6$ when terminal amine groups are attached to form cationic PTOBEE-$NH_2$. This was ascribed to the presence of the nitrogen atoms affecting auto-aggregation, that is, a compaction of one molecule by the introduction of positive charges or by formation of H-bonds with the neighbouring oxygen atoms, or by a combination of the two [19]. The apparent $R_g$ of PTOBEE-$NH_2$ calculated then (also from SANS data) at pH= 7, was 81 Å, assuming a rod-like nanostructure and 110 Å assuming a globular nanostructure. The latter figure is consistent with the value in Figure 5.

Work to better understand the origin of the pH-induced structural changes in these ChLCPs is in progress. However the results so far would seem to confirm the potential application of these new cationic ChLCP as endosomolytic polymers.

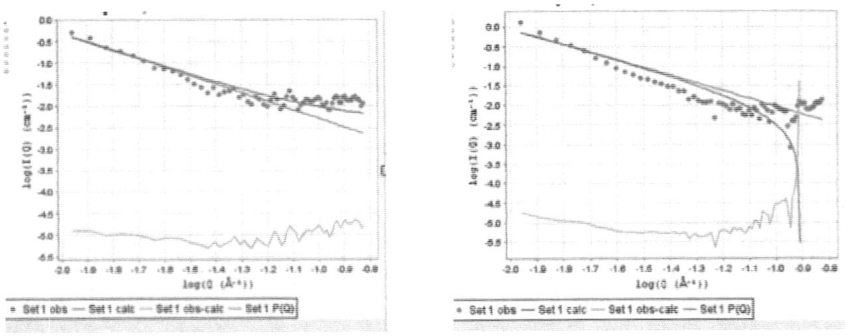


**(a)** **(b)**

**Figure 8.** Debye Gaussian coil model fits to the SANS data from PTOBDME-$NH_2$ at pH 1,3 (a) and pH 7,2 (b). The red circles represent the experimental data, the blue line is the model calculation. The green line is the difference between the two, superimposed on a yellow abcissa. The fitting was conducted with the FISH program (see text).

## 4. Conclusions

The SANS patterns of synthetic cationic cholesteric liquid-crystal polymers PTOBDME-Choline, PTOBDME-$NH_2$ and PTOBEE-$NH_2$, dispersed in $D_2O$ at 7 mg/ml, have been obtained at different pH values (= 1.3, 3.58, 5.58, 7.2), chosen to match the values in different cellular compartments (lysosome between pH 3.5 - 4, late endosome pH 5 - 6, early endosome pH 6 - 6.5 to cytosol, pH=7 - 7.4).

The apparent radius of gyration of the polymers, derived from the scattering data, proved to be highly sensitive to pH changes, initially increasing in size with increasing pH up to a maximum value, before subsequently decreasing once more at higher pH. The ChLCPs adopt rather different solution conformations to more conventional synthetic polymers, but the exact nature of the solution nanostructure and the explanation for the pH-sensitivity is, as yet, not fully understood.

The results shown here help indicate the potential of these new cationic ChLCPs as an endosomolytic release system in gene therapy, according to the hypothesis of "the proton sponge" where an endocytic pH change could be used as a signal to release a 'captive' biomacromolecule, such as DNA, in the correct intracellular compartment.

## 5. Acknowledgements

DR-M wishes to thank the MEC for a Scholarship of Specialization in International Organisms conducted at the ISIS Facility. STFC and ISIS are thanked for the provision of neutron beamtime.

## 6. References

[1] Anderson WF 1998 Human gene therapy *Nature* **392** (6679 Suppl) (25-30.

[2] Ibraheem D, Elaissari A, Fessi H Gene therapy and DNA delivery systems *Int J Pharm* 2014 **459**(1-2) 70-83.

[3] Griffiths PC, Paul A, Khayat Z, Wan KW, King SM, Grillo I, Schweins R, Ferruti P, Franchini J, and Duncan R 2004 Understanding the Mechanism of Action of Poly(amidoamine)s as Endosomolytic Polymers: Correlation of Physicochemical and Biological Properties *Biomacromolecules* **5** 1422-1427.

[4] Khayat Z, Griffiths PC, Grillo I, Heenan RK, King SM and Duncan R 2006 *Int. Journal of Pharmaceutics* **317** 175-186.

[5] Von Gersdorff K, Sanders NN, Vandenbroucke R, De Smedt SC, Wagner E, Ogris M The internalization route resulting in successful gene expression depends on both cell line and polyethylenimine polyplex type *Mol Ther.* 2006 **14**(5) 745-53.

[6] Behr, J P 1997 The Proton Sponge: a Trick to Enter Cells the Viruses Did Not Exploit *CHIMIA* **51** 34- 36.

[7] a) de Gennes PG and Hebd CR 1975 Macromolecules and Liquid Crystals: Reflections on Certain Research *Seances Acad. Sci. Ser. B* **281** 101;
b) http://plc.cwru.edu/tutorial/enhanced/files/lc/phase/phase.htm

[8] Ringsdorf H, Schlarb B, and Venzmer J 1988 Molecular Architecture and Function of Polymeric Oriented Systems: Models for the Study of Organization, Surface Recognition, and Dynamics of Biomembranes *Angew. Chem. Int. Ed. Engl.* **27** 113-158 ; c)

[9] Pérez-Méndez M and Marco C 1997 New synthesis, thermal properties and texture of cholesteric poly[ethyl ethylene 4,4'-(terephthaloyldioxy)dibenzoate] *Acta Polymerica* **48**, 502-506.

[10] Pérez-Méndez M and Marco Rocha C Preparing cholesteric liquid crystals - by adding acid di:chloride and butanediol to chloro-naphthalene, heating in nitrogen, decanting into toluene, etc Patent with n° EP1004650-A; WO9831771-A; WO9831771-A1; AU9854863-A; ES2125818-A1; ES2125818-B1; EP1004650-A1; US6165382-A; MX9906732-A1; JP2001513827-W; AU739076-B; EP1004650-B1; DE69824182-E.

[11] Fayos J, Sánchez-Cortés S, Marco C and Pérez-Méndez M 2001 Conformational analysis and molecular modeling of cholesteric liquid-crystal polyesters based on XRD, Raman and transition thermal analysis *J. Macromol. Sci.-Physics* ***B40***(3&4) 553-576.

[12] Sanchez-Cortes S, López-Ramírez M, Pérez-Méndez M and Blanch G 2010 Trans-cis Isomerization of Carotenoid Lycopene upon Complexation with Cholesteric Polyester Carriers Investigated by Raman Spectroscopy and Density Funtional Theory *Journal of Raman Spectroscopy* **41**, Issue 10, 1170-1177.

[13] Pérez-Méndez M, Fayos J and Mateo CR 1999 Self-assembly of cholesteric liquid crystal polyesters and their stereoselective interaction with liposomes of DMPC *Advances in Biochirality* Elsevier Science S. A. Chapter 24.

[14] Pérez-Méndez M, Areso S, Alarcón-Vaquero A, Elorza B and Malfois M 1998 Effect of polymer addition to lipid membranes as potential drug delivery systems. Structure and dynamics of their interaction *Annual Report EMBL* 372.

[15] Pérez-Méndez M, Marsal-Berenguel R and Funari, SS 2003 Structural Characterization of the Interaction Between Cholesteric Lquid-Crystal Polymers and Molecules of Biological Interest *HASYLAB Annual Report* 11150.

[16] Pérez Méndez M, Marsal Berenguel R and Sánchez-Cortés S 2003 New non-viral vectors based on biocompatible liquid-crystal polymers for the carriage and delivery of biomacromolecules and insoluble drugs as an strategy *Rev Oncol.* **4** Suppl 1 153.

[17] Pérez-Méndez M, Fayos J, Blanch GP and Sanchez Cortes S 2011 Biofunctionalization of Cholesteric Liquid-Crystal Helical Polymers. Nanocarriers *ENCYCLOPEDIA OF NANOSCIENCE AND NANOTECHNOLOGY* **11** 547-580, Edited by H. S. Nalwa, ACS. American Scientific Publishers, ISBN: 1-58883-160-4

[18] Pérez Méndez M Synthesis and characterization of new functionalized cholesteric liquid-crystalpolymers (to be published).

[19] Pérez Méndez M and Hammouda B 2013 SAXS and SANS investigation of synthetic cholesteric liquid-crystal polymers for biomedical applications *Journal of Materials Science and Engineering* **B 3** (2) 104-115.

[20] Perez Mendez M, Sanguino Otero J and Rodriguez Martinez D Structural characterization of cationic ChLCP/polynucleotide polyplexes by SAXS *To be published*

[21] a) http://www.isis.stfc.ac.uk/
b) Heenan, RK, Penfold J, King SM. *J. Appl. Crystallogr.* 1997 **30** 1140.
c) Heenan RK, King SM, Osborn R, Stanley HB 1989 COLETTE Users Guide *Rutherford Appleton Laboratory Report RAL*-89-128, (1989)
d) King SM, Heenan RK 1995 Using COLETTE *Rutherford Appleton Laboratory Report RAL-95-005.*
e) Wignall GD, Bates FS 1987 *J. Appl. Crystallogr.* **20**, 28.

[22] Zimm B H. The scattering of light and the radial distribution function of high polymer solutions *J. Chem. Phys.* 1948 **16** (12) 1093-1099 and 1099-1116.

[23] Aho AV, Hopcroft JE, Ullman JD. Data Structures and Algorithms, Addison-Wesley Longman Publishing Co., Inc. Boston, MA, USA ©**1983**.

[24] Benjaminsen RV, Mattebjerg MA, Henriksen JR, Moghimi SM, Andresen TL. The possible "proton sponge " effect of polyethylenimine (PEI) does not include change in lysosomal pH. *Mol. Ther.* 2013. **21**(1) 149-57.

[25] Midoux P, Pichon C, Yaouanc J J and Jaffrès PA Chemical vectors for gene delivery: a current review on polymers, peptides and lipids containing histidine or imidazole as nucleic acids carriers *Br. J. Pharmaco.l* 2009 **157**(2) 166-78

[26] Akinc A and Langer R Measuring the pH environment of DNA delivered using nonviral vectors: Implications for lysosomal trafficking 2002 *Biotechnology and Bioengineering* **78** 5 503-508.

[27] Sanjoh M, Miyata K, Christie RJ, Ishii T, Maeda Y, Pittella F, Hiki S, Nishiyama N and Kataoka K. Dual Environment-Responsive Polyplex Carriers for Enhanced Intracellular Delivery of Plasmid DNA *Biomacromolecules* 2012, **13** (11) 3641–3649.

[28] http://www.diamond.ac.uk/Beamlines/Soft-Condensed-Matter/small-angle/SAXS-Software/CCP13/FISH.html.

[29] Debye P. Molecular-weight determination by light scattering, *Journal of Physical and Colloid Chemistry*, 1947, **51**, :18-32.

[30] http://www.ccpsas.org

# AUTHOR INDEX


Albers, P W ........ 29
Appel, M ........ 63
Aramini, M ........ 71
Bellaire, D ........ 38
Brunelli, M ........ 71
Castañon, J Fernández ........ 20
Cavallari, C ........ 71
Chudoba, D ........ 8
Down, R B E ........ 57
Dudman, M P ........ 38
Elbert, J ........ 63
Evans, B E ........ 50
Ewings, R A ........ 38
Fernandez-Alonso, F ........ 20
Fojud, Z ........ 78
Frick, B ........ 63
Gaboardi, M ........ 71
Gallei, M ........ 63
Goodway, C M ........ 38, 50
Gorini, G ........ 20
Hetmanczyk, L ........ 8
Holderna-Natkaniec, K ........ 78
Ivanov, A ........ 1, 63, 71
Jimenéz-Ruiz, M ........ 1
Jiménez-Ruiz, M ........ 71
Jura, M ........ 38
Kazimirov, V Yu ........ 8
Keeping, J ........ 57
Kibble, M ........ 38
Kibble, M G ........ 50
King, S M ........ 94
Kirichek, O ........ 50, 57
Krawczyk, J ........ 8
Kulda, J ........ 1
Lennon, D ........ 29, 38
Major, R A ........ 57
Martinez, D Rodriguez ........ 94
Mazzani, M ........ 71
Méndez, S M Pérez ........ 94
Natkaniec, I ........ 8, 78
Parker, S F ........ 20, 29, 38
Peplinska, B ........ 78
Pinna, R S ........ 20
Pontiroli, D ........ 71
Ramirez-Cuesta, A J ........ 20, 29, 50, 57
Rehahn, M ........ 63
Riccò, M ........ 71
Rols, S ........ 71
Rudic, S ........ 20, 57
Sashin, I L ........ 8
Spehr, T L ........ 63
Stühn, B ........ 63
Taylor, J ........ 38
Tomkinson, J ........ 20
Tooze, R P ........ 38
Wakefield, S R ........ 38
Warringham, R ........ 38
Webb, P B ........ 38
Zalewski, S ........ 8

**Institute of Physics**
Dirac House, Temple Back
Bristol BS1 6BE UK